Treasure Hunting
with a Metal
Detector

In,
Around,
and
Under Water

by Steve Blount, Lisa Walker
and the editors of Pisces Books

 Pisces Books • New York

ILLUSTRATIONS

Cover Photographs: Margo Nelson; Joseph W. Jackson III of the *Wisconsin State Journal.*

Jan Higbee:10.
Robert Holland:C-8, C-10, C-15, C-16.
Joseph W. Jackson III:C-3 (top).
Margo Nelson:C-7 (bottom).
Rick Sammon:C-1 (bottom), C-6, C-7 (top), C-14 (top).
B. Sastre:C-1 (top), C-2, C-3 (bottom), C-4, C-5, C-ll, C-12, C-13, C-14 (bottom), 4, 7, 8, 9.

All equipment photographs courtesy of the various manufacturers.

Library of Congress Cataloging in Publication Data

Blount, Steve .
 Treasure hunting with a metal detector.

 Includes index.
 1. Treasure-trove. 2. Metal detectors.
I. Walker, Lisa. II. Pisces Books (Firm)
III. Title.
G525.B66   1986   622'.19   86-30357
ISBN 0-86636-049-2

Printed in Hong Kong

10 9 8 7 6 5 4 3 2 1

Staff

Publisher	**Herb Taylor**
Project Director	**Cora Taylor**
Executive Editor	**Virginia Christensen**
Editor	**Joanne Pfriender**
Art Director	**Richard Liu**
Production Associate	**Jeanette Forman**
Artist	**Dan Larkin**

"Treasure is all around you. You don't need a map to find it; all you have to do is look in the right place."

—**ROBERT LOUIS STEVENSON**
Treasure Island

Acknowledgments

Any book is a cooperative effort; this one perhaps more than most. The authors owe sincere thanks to Jim Lewellen of Fisher Research Laboratory for his generosity in supplying technical advice and information about metal detecting equipment. Special thanks are due also to Scott Mitchen, A.M. Van Fossen, Ron Houghton, Myron Higbee, Kevin Reilly, and Chester Morofsky. These treasure hunters graciously shared the experience they've gotten the hard way—they dug for it.

Perhaps the greatest dream of a treasure hunter is to discover and explore a sunken treasure ship. These historic coins were recovered from a Spanish galleon wrecked near Key West, Florida, more than 350 years ago.

Table of Contents

Introduction

What if you were told that, instead of painting the living room next weekend, by doing a little swimming and a little digging on Saturday, you could find enough gold to pay someone else to do the painting, and have enough cash left over for dinner in a fine restaurant on Sunday?

It's not only possible, it's probable. Real treasure is all around you— gold rings, valuable antique coins, jewelry, and historic artifacts are as close as the nearest beach, lake resort, or swimming hole. All you need to bring it home are a little common sense, some research, and some experience with an underwater metal detector.

But why in the water? Why not just find an old park or horse track and scan the dirt for goodies? Many treasure hunters do just that and reap good profits. But water sites are better for several reasons.

First, on land sites it is hard to define exactly where to search within a given area. At an old horse track the admission booth, grandstand, wagering windows, and concession areas would be good bets. But a lake, a stream, an ocean beach, or a canal bank offers much more definite possibilities. Around water, people's activities tend to be confined to a small area—a swimming beach, a rope swing, a pier or dock, a boat landing.

Second, on land, a coin or piece of jewelry dropped on the ground remains right there to be picked up by its owner or the next person who happens along. Coins and jewelry dropped in water are almost never recovered.

Third, cold water tends to make people lose things more often. Rings slip off swimmers' fingers. Diving off a pier causes a necklace clasp to snap. Why do people wear expensive jewelry while swimming or boating? Might as well ask why people walk out in front of moving cars every day of the week. They aren't thinking. Or they think the accident will happen to someone else.

And finally, the loose sand or mud bottoms common at beaches and lakes actually help the treasure hunter by covering up lost items, yet stabilizing them just below the surface where he can find them.

Convinced? Perhaps. But maybe getting your living room painted isn't enough incentive to invest money in equipment, and time and energy in searching. Would putting a down payment on a new house do it? Or paying cash for a new sports car? One treasure hunter found a man's sapphire ring appraised at $25,000 in his very first visit to Chesapeake Bay!

Now, here's some really good news. It's easy. You don't have to be a rocket scientist to use a metal detector. Metal detectors *are* complex, sensitive electronic instruments. The first models were cranky, sometimes unreliable and definitely not waterproof. But, some of today's detectors are *so* high-tech that they are simple to use. Most of them have no more than three knobs, one of them an on-off switch. And there are now several truly excellent waterproof detectors for searching in and around the water.

When you are ready to hit the water, you have two choices. You can wade, sweeping the detector across the sand from above, then digging with a long-handled scoop, or you can swim with a snorkel, a hookah airhose, or a scuba tank. How you search will depend a bit on the location, on the water conditions, and on your own comfort in the water. All of these methods have been used successfully by the treasure hunters discussed in this book to recover treasure ranging from Spanish coins— actual gold doubloons and silver pieces of eight—to high school class rings (in bulk, worth hundreds of dollars for their gold content) to diamond and sapphire jewelry.

Finding someone else's lost wedding ring is not like taking a chance by buying into a grab box at a garage sale. Unless you live in a particularly isolated area, there is a coin shop or a jewelry store near you that regularly buys and sells precious metals and antiques. You'll find a ready market for bulk gold and silver, coins, and pieces of jewelry.

And you'll find something else. An adventure. There's something distinctly magical about revealing a coin or a ring that's been lost for half a century or more. Who lost it, and when? What did they look like? What did they think, and how did they act? Was it a Rough Rider saying goodbye to a sweetheart before joining his regiment in the Spanish American War? A brass lock from an old stagecoach stop sends you scurrying in circles, looking for the payroll chest it once secured. A flapper or Prohibition-era dandy comes to life in your hands as you slowly rinse the mud off a gold-plated lipstick tube or a diamond stick-pin.

This is history in the making. Not the dry, stiff recitation of "great events" found in history textbooks. It's real history, real pieces of real peoples' lives.

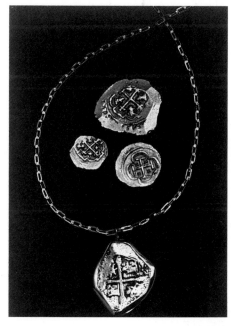

These Spanish treasure coins were recovered from a ship wrecked in 1715. Included here are a silver piece-of-eight that has been mounted as a pendant and 1-, 2-, and 4-escudo gold doubloons.

Chapter 1

A Flash of Gold

Treasure hunting has been romanticized beyond all reason. Books such as *The Buccaneers of America*, published in Amsterdam in 1684, Robert Louis Stevenson's *Treasure Island*, and *The Treasure of the Atocha* (recently published by Pisces Books) have rightly pointed out the wealth that lies just beneath the surface of many bodies of water. Unfortunately, too many readers ignore the most basic message of these books. They see the main characters as heroic, larger-than-life figures who are somehow cosmically predestined to take fate into their own hands and make it work for them. The fact is, as every one of those books shows, *anyone who has the desire can become a treasure hunter.*

Not everyone has the perseverance to spend 16 years looking for a lost Spanish galleon, as treasure hunter Mel Fisher did. But not everyone plays the treasure game for the same stakes. A lower investment of time, energy, and commitment can still yield substantial monetary returns. In fact, treasure hunters cover the whole range from casual weekend coin-shooters and dedicated hobbyists to small-stakes professionals and the super-hunters, like Fisher. In fact, many of the super-hunters started as occasional or weekend hobbyists.

Professional treasure hunters exploring unerwater wrecks use hand-held metal detectors at close range to find coins and other small pieces of metal. These pieces-of-eight were on their way to Spain when a hurricane destroyed the ship that was transporting them and left them on a reef in the Florida Keys.

Although most historic coins such as this are uncovered by professional salvors working undersea wrecks, amateur treasure hunters have also made some intriguing finds.

Kip Wagner

The man who started the great Florida rush for Spanish gold began his treasure hunting career very casually. One day in 1957, Kip Wagner, a retired building contractor, happened upon a coin while walking the beach near Sebastian Inlet, near Vero Beach, Florida. As soon as he had picked up the blackened, irregular-shaped object, Wagner knew it was something special. At home, the heavy oxide coating came off easily enough, and soon Wagner was holding a shiny lump of pure silver—almost a full ounce of it. The odd thing was the markings: a "crusaders," or Jerusalem, cross on one side and a four-quartered coat-of-arms on the other.

The coin intrigued Wagner. His favorite book was *Treasure Island*, and he began to picture a vast hoard of precious metal hidden somewhere on that beach. Wagner bought a primitive metal detector and used it to find more coins on the beach. When the trail of coins led him to the water's edge, he learned how to scuba dive and began hunting the shallow water just off the beach.

Wagner eventually sent several of the coins to Mendel Peterson, the curator of maritime history at the Smithsonian Institution in Washington. The coins Wagner had found were indeed part of a vast treasure hoard. He was on the trail of a great Spanish treasure fleet that sank in a hurricane off southeast Florida in 1715. The ships were carrying an estimated $1 million in gold and silver from Spain's New World colonies back to the mother country when they broke up on the shallow reefs off Sebastian Inlet.

Wagner's hobby soon became a full-time profession. He acquired some partners and a small boat. From this tiny beginning grew the first professional treasure company to operate in American waters—Real 8,

9

named for the silver eight *reale* coin that had sparked Wagner's interest. At one time or another, Real 8's partners have included other treasure hunters who would themselves become legends—Mel Fisher, Robert F. Marx, and Jack Haskins, among others. The company eventually recovered nearly $100 million worth of treasure from the 1715 fleet.

Sound like a boyhood dream come true? It is. But don't let the almost mythic dimensions of the finds or the stature of these treasure hunters fool you. Treasure is all around you, and you can find it. Believe that you can and start looking. Ordinary people do it every day.

Myron Higbee

There are few "rules" in treasure hunting. Kip Wagner started with his own two hands and a metal detector. Like him, Myron Higbee has proved that you don't have to work at it full time to turn a profit. And you don't have to live near the ocean.

Higbee began treasure hunting as an occasional hobby activity. It stayed a hobby until he found Rose Lake, in northern Idaho. Higbee's mother had a collection of old photographs that showed the lake had boasted several shore resorts up until the 1950s. "When I first started detecting, I was very casual about it. The pictures of Rose Lake showed me

These rings, medallions, and chains are the finds of one treasure hunter who hunts his favorite lakes even when the air temperature is 15 degrees.

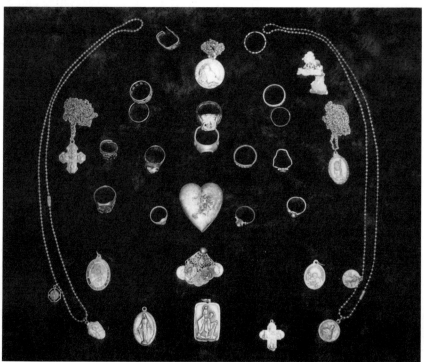

Using a long-handle scoop, waders, and a 1280-X Aquanaut, one treasure hunter found a $4,000 diamond ring in Idaho's Rose Lake.

it had great potential, so I began hunting with a land TR unit in a field overlooking the lake. When I dug up a Barber half-dollar, I realized this must have been an old picnic area. I found my first silver dollar in that field, and a number of dimes and quarters."

The allure of those first finds soon set off a full-fledged attack of the "Kip Wagner syndrome."

"What got me thinking about the water was that there used to be a dock at this picnic area. I knew there had to be a lot of coins around that old dock, so I converted my TR unit for shallow-water hunting. I found so much stuff that I wanted to hunt right out to the end of the dock, but the water was over my head. I signed up for a scuba diving course and bought an amphibious detector," Higbee says.

Like Wagner, Higbee followed a thin trail of silver out into the deep. And if his finds have been less spectacular than a sunken fleet of rich Spanish galleons, they've kept him in the water—and in the money—for three years. "I was out sweeping Rose Lake when a lady motored up in a small boat," he recalls. "She asked if my detector would find gold, and when I told her it would, she told me she'd lost a gold ring near her dock a few years back and asked if I'd look for it. The next morning, I went to her dock and waded out. I had spent about two hours pulling trash out of the sandy bottom when I got a really solid hit. I stuck my retriever in and came up with a gold ring that had the Rock of Gibraltar mounted on it—a huge diamond. The woman's husband was so grateful he insisted on giving me a bag containing 30 silver dollars, all minted in the 1800s, including some Carson City coins. Then he told me that the ring was appraised at more than $4,000."

Since that find, Higbee's casual interest in treasure hunting has taken a more serious bent—he now classifies himself as an "avid" hobbyist. Higbee hunts his favorite lakes even when the air temperature is 15 degrees Fahrenheit. "I make some terrific finds during the winter," he says. "I work around the old docks. In some lakes, the water level will drop as much as six or seven feet when the lake freezes, so I can hunt the mud left behind very easily. Also, I wade in and break up the ice out to where it's two inches thick. A lot fewer people hunt in the winter, and I prefer not to hunt in a crowd."

Scott Mitchen

Treasure does have a seductive influence on the imagination. Once you've pulled gold out of the ground, there's almost a compulsion to move on and up in the treasure hunting world. Scott Mitchen, who once hunted only casually, is now a full-time treasure hunter. Like Higbee, Scott Mitchen lives thousands of miles from the ocean, in Milton, Wisconsin. The state's hundreds of lakes are a fertile field for exploration. In fact, Mitchen rarely hunts areas more than two hours from his home. Even in so restricted a radius, Mitchen has not run out of productive sites: "I go back to the same site again and again. I like going into areas that literally hundreds of other people have worked, both waders and divers. I dive almost exclusively, even in water that's only a few feet deep. Sometimes the top of my tank is bobbing out of the water. I can recover gold that's beyond the range of the metal detector by actually digging down into the bottom. If it's a spot I'm sure of, I'll just move the bottom and work deeper and deeper—far beyond the range of waders."

Mitchen's rewards? "In one of these areas I brought out a gold ring with a full carat diamond in the center flanked by four smaller diamonds. Just recently I went out for a couple of hours, just poking around really. I came back with a ring that was set with a ruby and four diamonds. I stopped at a bar to celebrate and ran into an old girlfriend and her husband. Well, she put the ring on and never took it off. They offered me $500 for the ring. I knew it was worth more, but it was kind of neat to find something an old friend liked so much, so I sold it. Later, she told me she'd had it appraised at over $1,000. Going out for a couple of hours and coming back $500 richer is definitely nice."

Mitchen searches old ferry crossings, stagecoach stops, lake resorts, and old swimming beaches. His most exciting find, he says, really doesn't have much monetary value. "There's an old stagecoach stop called Dutch Mills near Milton. Although I've worked it pretty heavily, I'm still pulling coins out. One day, though, I found an ancient brass padlock—the kind

Scuba gear enables treasure hunters to explore beyond the water's edge and often gives them access to areas rich in treasure.

they used on jail cells in the 1800s—lying in the silt. The key was still in it! I picked it up and took it home as a souvenir. I soaked it in some WD-40 and, while showing it to a neighbor, I inadvertently turned the key and it popped open! It still worked. My grandfather is a locksmith and gunsmith, and when I told him about it on the phone, he got really excited. I looked around the area where I found the lock; I thought maybe there was a strongbox or something down there. I did find another brass padlock, and a key lying right next to it."

Aside from selling a few of his better pieces to friends, Mitchen shares his wealth in other ways: He teaches courses in underwater metal detecting. "I'm able to teach people how to use the machines quickly. Two guys who took my course went out the day after we finished, and each of them found a gold ring."

Even after making the transition to full-time treasure hunting, Mitchen feels compelled to buy in a little deeper still. He now has his sights set on the Great Lakes. "There are so many shipwrecks in the Lakes, and more than a few were carrying gold or silver shipments. I'm teaming up with a guy who has a charter boat. During the week, we'll work the shipwrecks and on the weekends, he's going to run dive trips to the wrecks we find."

A.M. Van Fossen

The next step for someone who truly believes in treasure—and has proved to himself that it's not just a fantasy—is to go into true marine salvage. Van Fossen is someone who believes. In spring 1986, Van Fossen and his diving partner, Ron Houghton, spent a few days hunting with super-hunters Jack Haskins and Johnny Berrier aboard their salvage boat, the *Wasp*. Their haul for what was actually something of a pleasure excursion included a number of silver *reales*. The wreck they were working was part of another Spanish treasure fleet, this one lost in 1733. Van Fossen and Houghton are no strangers to the ultimate treasure experience—working a Spanish galleon.

Van Fossen and Houghton are based in Houston, where Van Fossen operates Research and Recovery International. A tireless and knowledge-able researcher, Van Fossen provides investment partnerships with tax-sheltered treasure expeditions. He researches wrecks in archives all over the U.S. and Europe, looking for clues to their probable location and their cargo. Once a wreck is identified and known to have been carrying treasure, an investment group is offered the chance to fund a salvage operation on the vessel. Van Fossen's firm is paid for research done to identify and define the search area where the wreck should be located. The investment group pledges funds for a multi-year search and recovery effort.

Research and Recovery then goes out and actually finds the wreck. After the identity of the wreck is confirmed, a salvage operation is mounted using contract commercial divers—divers with deep-diving experience on oil rigs.

Their most exotic find? On one expedition Fossen and Houghton came up with a couple of Spanish three-*reale* coins. What makes them so unusual is that none of the scholars of Spanish colonial coinage who have examined the coins has ever seen or heard of a three-*reale* coin. According to the official records of the Spanish mints, no three-*reale* coins were ever minted. But Van Fossen and Houghton know they were.

Already participants at the headiest level of professional treasure hunting, Van Fossen and Ron Houghton are also subject to that seductive lure that has pulled Higbee and Mitchen further and further away from the work-a-day world. They have a contract to salvage the *Central America*, a vessel belonging to the U.S. Postal Service that sank in 1857. The payoff? The *Central America* may be the richest ship to sink in American waters. It is thought to have been carrying a cargo of golden "eagles"—gold coins—worth $200 million.

Ordinary People

There's nothing magical about any of these treasure hunters. Their one distinguishing characteristic is a determination to do what suits them, and not to listen to the cynics. These ordinary people have reached into the realm of legend and pulled out a small, and valuable, piece for themselves. While it is hard to tell precisely how much the *reales* recovered by Van Fossen and Ron Houghton are worth—other than a substantial amount— the finds recorded by others seem a little pale by comparison. Yet the more mundane recoveries can be very valuable.

Wallace Chandler, a dedicated hunter for nearly two decades, made his first trip to the Chesapeake Bay last year. With his two partners, Chandler hunted the usual tourist beaches, and the impressive haul of silver coins and rings was more than enough to keep them interested. An older gentleman wandered by, and on hearing about their expedition, suggested they try an abandoned swimming beach just a bit further along. The three fully expected this hot tip to turn out to be a false alarm.

Arriving at the location described by the man, they found not a beach, but a grassy slope running down to a seawall. One of the partners did not even bother to unpack his gear. However, Chandler and his other friend scrambled down over the seawall and began wading. Chandler's first scoop brought up a Mercury dime—not exactly a world-class find. The next two signals proved to be large lead fishing sinkers. Chandler was ready to pack it in when a fourth loud signal sounded in his headphones. Expecting another lead sinker, he casually dug the retriever into the bottom and pulled up a man's 24-karat gold ring set with a 7.65 carat deep blue sapphire. The value? Chandler has had the piece appraised at more than $25,000.

Chapter 2

Get Equipped!

All serious treasure hunters agree on one thing: good equipment is the key to making good finds. The most important piece of gear, of course, is the metal detector. The chart in this chapter lists the most popular waterproof detectors currently on the market. Buy the best one you can afford, but remember that this is not an expense, but an investment. In addition, there are a number of relatively inexpensive accessories that will make your hunting more pleasurable and more profitable—these are described in the section following the chart.

Because metals conduct electricity, they will cause changes in any electromagnetic field that comes into contact with them. Among the first metal detectors were devices that simply read the earth's own magnetic field and measured any fluctuations in the strength of that field as the detector was moved. These devices, known as *magnetometers*, are useful for finding big chunks of ferrous (iron-bearing) metal, such as ship's hulls, anchors, and engine blocks. Most of the Spanish treasure wrecks salvaged in Florida have been found using magnetometers. Unfortunately, magnetometers are not effective for finding small, non-ferrous items (metals without iron in them, such as silver or gold).

Magnetometers, such as this one are towed behind a boat to find very large iron or steel objects. Magnetometers and metal detectors work on the same principle—that a metal object will create variations in an electromagnetic field.

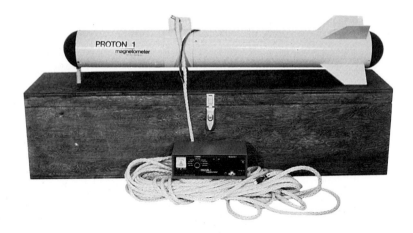

Anchors, cannons, and other large metal objects may be the first indication of the presence of a shipwreck. Once large objects are located, a search for smaller, harder to detect objects can begin.

On land, research can begin. Here a search team, being filmed for a documentary, begins examining the ruins of an old Spanish pearl fishing camp. The site, off the north coast of Venezuela, was abandoned in the 18th Century.

Modern Finds

This was the find on a beach after a full day's work in winter. The beach, popular in summer, was battered the week before the search by a storm that moved tons of sand.

By visiting areas where new divers train it is possible to find weights, knives, and other equipment lost by student divers. A quarry in Pennsylvania was the site of this find after a weekend in which many divers did their first open water dives.

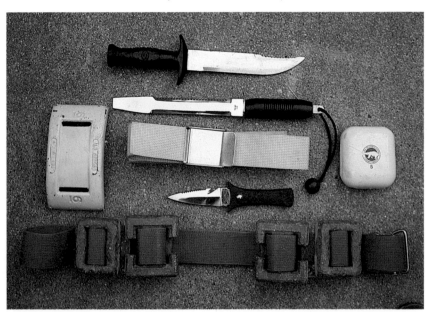

A successful treasure hunter flashes the jewelry, old coins, and artifacts he recovered from Lake Superior near his home in Wisconsin.

These momentos were found in the sand surrounding the wreck of a warship that sank off the south shore of Long Island. Before recovering and handling munitions, whatever the age, it is a good idea to obtain expert advice.

Historical Treasure

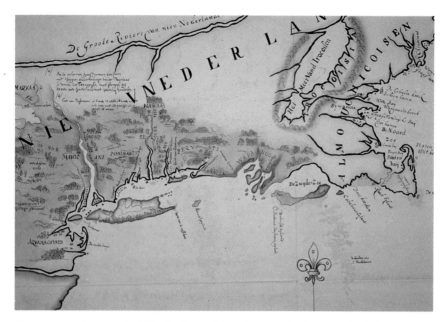

Old maps can be a great aid in finding suitable places to begin a search. This one, however, showed too big an area to be useful.

Searching often yields interesting non-metallic objects. The clay pipe and old oyster shells were found close to each other at what may have been the remains of a water-front tavern.

In the cold murky waters of Long Island Sound is the wreck of an 18th Century British warship that was burned to the waterline to prevent it from falling into the hands of the rebels. Though only the hull timbers remain, the surrounding sand has yielded cannons, hand weapons, and some coins.

Though the wooden parts of this pistol had almost completely rotted away, enough pieces remained to identify it as an 18th Century weapon probably of European origin.

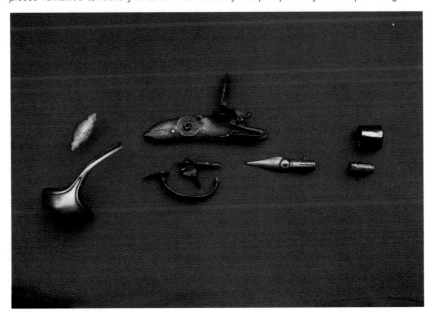

Small hand-held detectors can be especially useful when working in water that is murky. Having the search coil close also makes it easy for one person to search and probe at the same time.

This detector incorporates an arm brace in the handle to ease the strain on forearm muscles.

This detector has an arm brace and an adjustable shaft that permits the search coil to be moved closer or farther away so that the ideal balance position can be found.

Spanish Treasure

Metal detectors have been used extensively in the recovery of treasure and artifacts from the wreck of *La Nuestra Senora de Atocha* off Key West, Florida. Here, a diver has recovered a silver candlestick.

This growth-encrusted sword was found in shallow water off the east coast of Florida. On that coast are the wrecks of many Spanish treasure galleons. The ships passed close to the coast on their way back to Spain and they frequently sailed at the height of the hurricane season.

Careful searching (and re-searching) can prove quite productive. This diver is working around the remains of a Spanish treasure galleon.

One musket ball just like these convinced treasure hunter Mel Fisher that he had located the site of the wreck of the Atocha. He was correct, but it took many years and millions of dollars to find the main portion of the wreck.

Coins can help identify shipwrecks. These three Spanish coins are from three different eras. Left to right: 18th Century ''cob,'' 17th Century ''piece of eight,'' 19th Century coin.

Research

Old prints can be extremely helpful in identifying objects recovered from the sea. Though the cannon and weapons in this print are depicted in a general way, the navigation instrument being used could be an aid in dating or identifying a similar piece.

Drawings, paintings, and prints of old sailing ships are great aids in the identification of shipwrecks. Once construction details are known, much can be inferred from the remains of the ship even if only a small portion of it exists.

By examining old prints carefully clues to shipwreck locations can sometimes be found. The wreckage in the foreground of this print and its relative position to the pier could be an indication that the ship had sunk there. If the remains of the pier could be located today, an exploratory dive or two might be worthwhile.

Methods

Working as a team can make searching for objects move at a rapid pace. One diver can manipulate the detector while the other can probe for goodies. A non-metallic probe is essential and care must be taken to avoid readings from rings or watches worn by the person with the probe.

Using the non-metallic probes shown here, two divers recovered the gold objects from a beach-front country club in the Philippines.

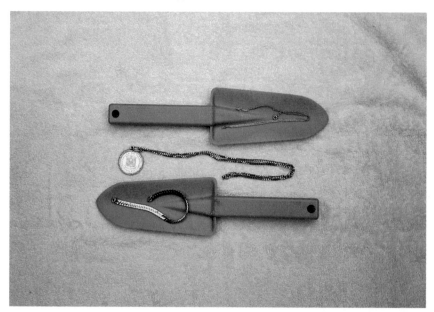

Gold coins are what every diver/treasure hunter hopes to find. These were recovered from the wreck of the Atocha.

Mel Fisher is probably the world's foremost treasure hunter. For over 16 years he searched for and ultimately found a Spanish treasure galleon with over $400 million in gold, silver, and jewelry. The story is told in Treasure of the Atocha, *by Duncan Mathewson, the archaeologist who joined the search team and helped save and preserve many of the objects recovered.*

In the late 1920s, Dr. Gerhard Fisher obtained the first patent issued on aircraft radio direction finders. He observed that aircraft pilots using this device were finding errors in their bearings when metal was present between the transmitter and receiver, or whenever they were flying over highly conductive, mineralized areas. Dr. Fisher concluded that the same principles at work in creating these errors could be used to develop a portable prospecting instrument for detecting small buried objects and ore deposits. After founding Fisher Research Laboratory in 1931, Fisher and four employees began producing the *Metallascope*, a rugged, perhaps ungainly but easy-to-use metal detector. In 1937, Dr. Fisher was granted the first metal detector patent.

Even though they don't contain iron (which causes massive fluctuations in the earth's magnetic field), gold, silver, aluminum, and other metals will cause very faint disruptions in an electromagnetic field. Manufacturers have developed several new types of instruments to detect these slight disruptions. Most of them not only receive—or read—the electromagnetic signals, they send them out—or transmit them—as well. The frequency of the signal in kilohertz (number of wave crests per second, in thousands) and the size of the search coil are just two of the factors that affect the kinds of metals a detector will "read," the type of soil or water it will work in, and how deep into the earth it will read.

One If By Land, Two If By Sea

There are dry-land detectors, water detectors, and amphibious detectors designed to work on land or in water. Which is best?

Dry-land detectors can be "converted" for use in water, that is, the electronics can be encased in Lexan or some other type of plastic to help waterproof them. Unless this waterproofing is done by an expert, it won't last long. When it comes to electric gizmos, water is an insidious substance. It will find some microscopic pore and migrate immediately to the most critical part of the device, shorting it out. Your dreams can literally disappear in a puff of foul-smelling smoke.

For hunting in and around water, most professionals recommend an amphibious detector. These units are fully waterproof, generally to a depth of 45 to 75 meters (150 to 250 feet). Even if you plan to search no deeper than your ankles, you would be better off with a fully waterproof unit. At any time you could stub your toe on a rock and drown those expensive integrated circuits.

Amphibious metal detectors are waterproofed the same way that cameras or other electronic gear are. The search coil, which transmits and receives the electrical signals, is encased in a waterproof resin. Then a housing made of a durable, non-corroding material such as Lexan, PVC, aluminum, or steel is used to encase the electronics.

In general, the fewer openings and fittings on a watertight case, the better. Most manufacturers seal the electronics compartments at the factory, which keeps them safe from leaks that may penetrate other parts

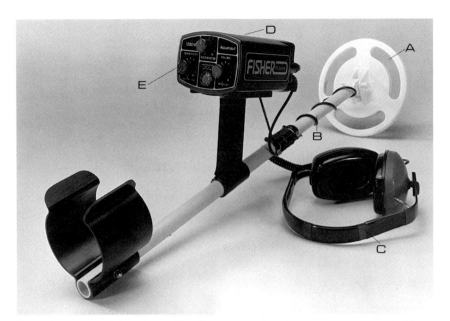

Some detectors are fully submersible. The search coil and electronics are protected by a waterproof casing. The parts of the detector shown here include: the serach coil (A), an adjustable shaft (B), headphones (C), the control head (D), and the control panel (E).

of the case. This is a good feature. Look over the cables and other attachments to the case. Are the fittings durable and tight? A hardwired connection is usually more reliable in water than a detachable one.

If the unit runs on replaceable batteries, access to the battery compartment is through a cover that is generally sealed with an O-ring—a round, circular rubber gasket. When greased with silicone and compressed slightly, this gasket will make the fitting between the cover and the body of the unit waterproof. Before buying any amphibious model, check the housing, seals, and O-ring seats. A smooth housing indicates good workmanship and careful casting—essential to long life in a watertight housing. Also, the O-ring seats must be smooth and free of pits or gouges that might allow water to seep in under the gasket. After you buy, the seals must be carefully attended to each time you use the detector and whenever you change the batteries. See Chapter 4 for easy maintenance tips.

Mixed Signals

Aside from being classified as land-only or amphibious machines, detectors are grouped by the kind of electrical signal used to discern the targets: RF-TR, VLF-TR, PI, and VF-TR.

RF-TR. The acronym stands for *Radio Frequency Transmitter-Receiver*. These detectors have two search coils contained in a single housing—one a transmitter and one a receiver. The transmitter coil generates a signal usually around 100 kHz. The signal forms a symmetrical electromagnetic field around the coil. Metal objects will alter the shape and intensity of the field, and thus be detected by the receiver coil. RF-TR units are relatively unsophisticated and inexpensive. However, their ability to read metal targets is severely limited by the presence of dissolved salt in water (such as in the ocean) or of minerals in the bottom covering. RF-TR units are best suited to fresh water and areas where there are few minerals present in the soil.

VLF-TR. *Very Low Frequency Transmitter-Receiver* units work the same way as RF-TRs, except their signal is at a much lower frequency, generally 3 kHz to 30 kHz. VLF-TRs can be "tuned" to ignore dissolved salt and moderate amounts of ground minerals. Some models can also reject signals from small pieces of tin foil, pop tops, and other trash. While VLF-TRs can be used in salt water, the combined presence of dissolved salt, ground minerals, and junk often produces false readings. Like RF-TRs, they are most useful in fresh water or on land.

PI. *Pulse Induction* detectors have a single search coil that emits pulses—short bursts of electromagnetic waves—100 to 3,000 times per second, depending on the model. Between pulses, the coil detects currents induced into buried metal objects by the pulses.

PI units are immune to the effects of salt water and heavily mineralized ground. For this reason, they were the first type of detectors successfully used in ocean searches. Many of the exquisite small items found on the Spanish shipwreck *Nuestra Señora de Atocha*, a galleon that yielded nearly a half-billion dollars worth of treasure to professional Mel Fisher, were found with pulse induction machines.

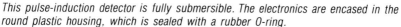

This pulse-induction detector is fully submersible. The electronics are encased in the round plastic housing, which is sealed with a rubber O-ring.

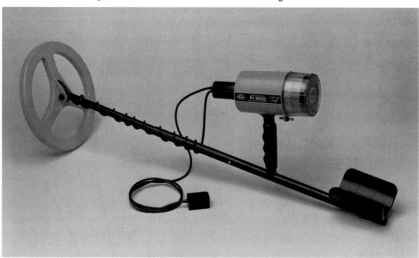

Underwater/Amphibious Metal Detectors

Model	Type	Search Frequency	Uses	Tuning Control	Discrimination Control	Sensitivity Control
Fisher Research 1280-X Aquanaut	VF-TR	2.4 kHz	Salt, Fresh	No (Automatic)	Yes	Yes
J.W. Fishers Mark I Audio	VLF-TR	5 kHz	Salt, Fresh	No	No	Yes
J.W. Fishers Pulse 6	PI	100 Pulse/Sec	Salt, Fresh, Blk Sand	No	No	Yes
J.W. Fishers Pulse 8	PI	N/AV	Salt, Fresh, Blk Sand	No	No	Yes
Garrett XL 200 Sea Hunter	PI	115 Pulse/Sec	Salt, Fresh, Blk Sand	Yes	Yes	Yes
Garrett XL 500 Sea Hunter	PI	115 Pulse/Sec	Salt, Fresh, Blk Sand	Yes	Yes	Yes
Gilbert Treasure Ray Aqua Pro TR	RF-TR	85 kHz	Fresh	Yes	No	Internal
Gilbert Treasure Ray Aqua Pro T/R Convertible	RF-TR	85 kHz	Fresh	Yes	No	Internal
Gilbert Treasure Ray Aqua Pro Pulse Convertible	PI	120 Hz	Salt, Fresh, Blk Sand	Yes	No	No
Turtle TR	VLF-TR	5.25 kHz	Salt, Fresh	Yes	Yes	No
Turtle VLF/TR	VLF-TR	5.25 kHz	Salt, Fresh	Yes	Yes	No
White's PI 1000	PI	3.2 kHz	Salt, Fresh, Blk Sand	Yes	No	No

ABBREVIATIONS

Type of Detector

PI—Pulse Induction
RF-TR—Radio Frequency Transmitter-Receiver
VF-TR—Voice Frequency Transmitter-Receiver
VLF-TR—Very Low Frequency Transmitter-Receiver

Batteries

AA-Repl—1.5 volt AA replaceable batteries
9v-Repl—9 volt replaceable batteries
Ni-Cad—Nickel-Cadmium rechargeable batteries

Other Features	Depth Selector	Audio Signal	Headphone Type	Coil Diameter	Battery	Estimated Price
Silent, No Threshold Operation	No	400 Hz, Volume Adjustable	Dual U/W Ear Cups	8-inch, Fixed, Shielded	AA-Repl or Ni-Cad	$649.95
Meter, Threshold Operation	No	0–15Hz	Single Ear Cup	10-inch, Fixed	9v-Repl	$359.00
Meter, Threshold Operation	No	0–15 Hz	Single Ear Cup	10-inch, Fixed	Ni-Cad	$595.00
Meter, Threshold Operation	No	Variable	Single Ear Cup	10-inch, Fixed	Ni-Cad	$795.00
Meter, Threshold Operation	No	430 Hz	Dual Land Phones or Vibrator	8-inch, Fixed, Shielded	Ni-Cad	$599.95
Meter, Threshold Operation	Yes	430 Hz	Dual U/W Ear Cups or Vibrator	8-inch, Interchangeable, Shielded	Ni-Cad	$799.95
Meter, Threshold Operation	No	400 Hz, Volume Int Adj	Single Ear Cup	8-inch, Fixed, Shielded	AA-Repl	$290.00
Meter, Threshold Operation	No	400 Hz, Volume Int Adj	Dual Ear Cups	8-inch, Interchangeable, Shielded	AA-Repl	$390.00
Meter, Threshold Operation	No	120 Hz	Dual Ear Cups	8-inch, Fixed, Shielded	C-Repl	$599.95
Threshold Operation	No	400 Hz, Volume Adjustable	Dual Ear Cups	8-inch, Fixed	9v-Repl	380.00
Threshold Operation	No	400 Hz, Volume Adjustable	Dual Ear Cups	8-inch, Fixed	9v-Repl	$540.00
Threshold Operation	No	N/AV	Vibrator	11-inch, Fixed, Shielded	AA-Repl or Ni-Cad	$500.00

Other Abbreviations

Blk Sand—Black sand, magnetized sand
Fresh—Fresh Water
Hz—Hertz (cycles per second)
Int Adj—Internally adjustable
kHz—Kilohertz (thousands of cycles per second)
Land Phones—Non-submersible headphones
Salt—Salt water
Sec—Second
Threshold Operation—Instrument must be tuned for faint audio ''threshold'' tone
U/W—Underwater submersible headphones

21

Currently, PI detectors still have an advantage over other kinds of units in terms of the depth of bottom material they will "read" through. A professional-grade PI detector may hit a coin as small as a dime under more than two feet of sand!

The current generation of pulse induction equipment, however, suffers from some shortcomings, too. These detectors are extremely sensitive to iron. While this is a boon if you are looking for a cannon buried deep in the bottom of the ocean, it can be annoying when nails and other debris register more strongly than silver or gold coins. What's more, the machines respond slowly to targets; the operator has to move the search coil very slowly, which makes it a bit harder to pinpoint the exact location of a target. Perhaps the most inconvenient aspect of PI detectors is that they must be "tuned" to the mineralization of the area being searched. Holding the coil a fixed distance above the ground, the machine takes a reading of the basic background electrical conductivity of the soil. This effectively "cancels" the machine's response to the soil and helps eliminate false readings caused by ground minerals. However, you'll get a false reading if the coil is not held at precisely the same height above the ground during the search as it was when it was tuned. This is hard to do, particularly in water. If you are wading, you may not be able to see the coil clearly enough to ensure that it is swinging at the right height. If you're snorkeling or diving, waves, surge, or an uneven bottom may make this impractical.

VF-TR. A new kind of unit has been introduced that combines the effectiveness of the PI in salt water or highly mineralized areas with the ease of use of a TR detector. These units are called *Voice Frequency Transmitter Receivers*. Like other TRs, they emit a constant signal, but at a much lower frequency—about 2 kHz. The VF-TR models now available automatically adjust to the ground mineralization and, unlike PIs, do not need to be tuned. You can vary the height of the search coil above the bottom without false signals. While VF-TR models are not as sensitive to deep targets as a PI, seasoned hunters say they can push the coil of a VF-TR unit into loose sand or mud, reading targets that most detectors won't reach.

VF-TR detectors do have one limitation—they cannot be used in high concentrations of "black sand." Black sand is highly magnetized sand (also called *magnetite*) that is found in some ocean beach areas. It is common along the extreme northwest and extreme northeast coasts of the United States, for example, although it is less common elsewhere.

Hunter's Choice. If you will be hunting exclusively in fresh water, you can choose from a wide variety of machines at a variety of prices. As with most things, you get what you pay for in a metal detector. There are no "runaway bargains." Given the fact you'll be devoting a fair amount of time to finding good search areas and then sweeping them, your payback in terms of treasure found will be a lot higher with a good-quality machine: one that can "see" a small target through at least six inches of sand, can be tuned to ignore signals from magnetite and other background signals, can discriminate between tin foil and treasure, and is reasonably easy to use.

If you will be hunting in fresh *and* salt water, there are only two real choices—a pulse induction or a voice-frequency TR detector. Either machine will operate well in salt or fresh water, will tune out background signals, and will give you good "reach" down into the bottom. The pulse induction has an advantage in working highly mineralized areas, while the VF-TR is a great deal easier to use, especially on an uneven bottom, in currents, or in surge.

Operating Manual

Tuning. The control panel of the first metal detectors looked like the cockpit of a jumbo jet, with an amazing variety of knobs and meters. Over the years, the number of controls you have to manipulate to find treasure has been reduced. You want a detector with a few controls to help you customize the machine's response to the area you're searching. On any machine except a voice-frequency TR, you need some type of ground balancing and/or tuning control to eliminate false signals caused by background magnetism.

Discrimination. Better detectors have the ability to reject some signals. A discrimination control will allow you to "tune out" faint signals, such as those from tin foil, nails, and other trash. Professional treasure hunters usually hunt with their discrimination turned up just high enough to keep from getting a signal from a gum wrapper. By turning up the discrimination, they can eliminate signals from pop-tops as well, but at that level, the machine may not read small gold rings. Your machine should have a discrimination control, one that is continuously variable from no discrimination at all to a setting that effectively tunes out most small pieces of trash. To set your discrimination, you might try burying a variety of objects in your backyard. Get some nickels, silver coins, rings, pop-tops, nails, and other trash. Bury the objects at 5 cm (two inches) and read them, then bury them at 10 cm (four inches) and at 20 cm (eight inches). Note that, as you turn the discrimination up, you have trouble reading even larger objects buried deep and the nickels do not register at all. Find a setting that rejects foil, but doesn't reject rings or pop-tops.

Depth/Sensitivity. A depth or sensitivity control is useful as well. Most of the time, you will want to search as deep as possible. With the control, however, when you get a signal, you can try decreasing the depth to determine whether you've got a small object near the surface or a bigger object down deep. A chunk of iron under four feet of sand may give the same strength of signal as a quarter at six inches. The depth/sensitivity control can help eliminate some of these signals and give you an idea about how deep to dig.

Volume. Some detectors allow you to adjust the volume of the signals coming through the headset. A signal that is almost uncomfortably loud in one search area may not be loud enough to be distinct in another. Some detectors also allow you to adjust the pitch of the signal, how high or low it sounds. That's nice, but it's not a necessity unless your hearing is

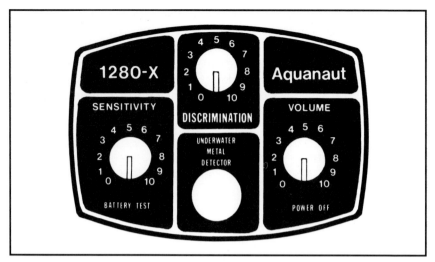

The controls shown on this panel of a VF-TR detector can be used to fine tune the unit to give the best performance under a wide variety of field conditions. The discrimination control rejects low-level signals, signals from minerals in the ground or those from base metals, such as aluminum. The sensitivity control can be used to limit how deep the signal penetrates into the ground, to determine, for instance, whether a target is a large object buried deep or a smaller one near the surface.

impaired at some sound frequencies. If you have some difficulty hearing as a result of years of diving, check the signal tone before buying to ensure that it is one you can hear well. The signal tone of most detectors is in the 400-Hz range.

Meter. Some detectors come equipped with visual meters that show the strength of the signal. The meters give you some idea of what is being read—a pop-top or a silver coin, for instance—and an indication of the depth of the object. While this is certainly useful, it's far from essential as you will want to dig for every object you identify, even if it is junk. The signals from junk close to the surface tend to mask the weaker signals from valuables buried a bit deeper. Also, where junk accumulates, coins and jewelry may accumulate as well.

Headphones. Headphones are supplied with almost all units. Unlike Walkman headphones, these are waterproof and can be fully submerged. They may be *single cup* or *dual cup*. With single cup headphones the signal comes through only one ear, while dual cup headphones pipe the signal to both ears. Your brain was designed to hear things out of both ears (which is why it likes stereo music so much better than monaural). Part of the art of reading signals is to learn to distinguish, by the characteristics of the detector's audio signal, what kind of object is being detected—is the signal short and brash, or smooth and mellow? Short, clipped signals usually mean iron or other trash, while coins and gold give a much "smoother" audio response. At first, you won't be able to distinguish one signal from another. But if you hear the signals time after time, you will begin to recognize the "tune" of a gold ring and the slightly different "tune" of a pop-top. Dual-cup earphones aren't essential, but they can help you learn the tunes more quickly.

Some manufacturers supply small mechanical vibrators instead of headphones. These vibrators emit an audible click and a physical pulse. When held against your temple by a mask strap or headband, you can actually feel the signal and hear it as well. This works particularly well underwater. However, you can't read the signal from a vibrator as you can the audio tone from headphones.

Keeping headphones on under water can be a bit trying. They tend to get knocked about, especially if you're working in a current or a surge. Although everyone seems to have his own solution, the most popular method is to remove the earcups from the plastic headband and push them up under a wetsuit hood. This keeps them firmly in place. **However, when you are diving, do not put the earphones over your ears at the surface and then dive. Wait until you are at your maximum depth before covering your ears with the cups**. If your ears are covered by tight-fitting headphones, the water cannot get to your eardrums. You will not be equalizing the pressure in your inner ear with that of the water outside the earcups. Your inner ear will"think"it is surrounded by air at 14.7 pounds per square inch. If the earcup should get dislodged while you are down at 32 feet, for example, your eardrum would be confronted with a sudden doubling of pressure. This could lead to several highly unpleasant and painful otolaryngological events, such as a ruptured eardrum. If you will be in water less than six feet deep, this won't be a problem. But, to be safe, always equalize your ears at your working depth before covering them with the earcups. It's a good idea to take the earcups off before heading to the surface, as well. Also, **do not use the little earplug-type earphones that fit within the ear canal**. Aside from preventing you from equalizing your ear properly, the water pressure could drive them deeply into your ear canal.

Shaft. Aside from the headphones, the two primary components of the detector are its search coil and control box. Both are mounted to a shaft. The shaft is what holds the coil away from your body. Most are made of fiberglass and the length can be adjusted easily. The control box is normally mounted near the top of the shaft. Some hunters prefer to have the control box clipped to their belts. On many models, the control box and coil can be removed from the shaft and mounted separately. This is useful, particularly if you want to mount the coil on a shorter shaft for easier handling in the water.

Coils. The size of the search coil affects the depth of penetration of the signal and the ability to "pinpoint" the location of a target. A bigger coil penetrates deeper, but isn't as accurate in showing the precise location of the metal. Most units are supplied with a coil about 20 cm (eight inches) to 25 cm (ten inches) in diameter. These give a good balance between depth of detection and ability to pinpoint. Larger or smaller search coils are available for some detectors. The smaller coils might be useful for probing in between and under rocks at a wreck site, for example. They are very precise in pinpointing objects, but only read just below the surface.

Electro-static shielding helps protect the coil from false signals caused by certain ground conditions.

Power. Detectors may be powered by either replaceable carbon or alkaline batteries or by rechargeable nickel-cadmium (ni-cad) batteries. The primary advantage of ni-cads is that, over the life of a unit, it is cheaper to buy them than it is to buy replaceable batteries. The detectors use so little current, however, that replaceable batteries last quite a while—up to 40 or 50 hours of use. The best ni-cads will only power a unit for 8 to 10 hours of hunting, requiring a recharge after every use. This may or may not be convenient for you. One thing to remember about ni-cads is that they can develop a "memory" if charged improperly. If you used the ni-cads for two hours each day, recharging them after every use, eventually they would lose their ability to run the unit for more than two hours. If you buy a unit with ni-cads, avoid giving them short "booster" charges. Instead, let the batteries run down close to zero each time before recharging.

Other Equipment

You'll need more than a detector to be an effective searcher. The gear varies, of course, depending on whether you'll be wading, snorkeling, or diving.

A wetsuit offers the very best protection against cold water and abrasion. Any jacket shown in the top row here can be used with any item in the bottom row to achieve a certain degree of warmth.

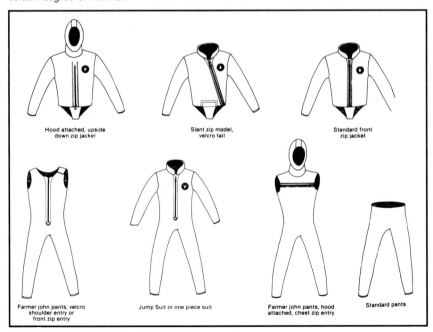

Hood attached, upside down zip jacket

Slant zip model, velcro tail

Standard front zip jacket

Farmer john pants, velcro shoulder entry or front zip entry

Jump Suit or one piece suit

Farmer john pants, hood attached, chest zip entry

Standard pants

Thermal Protection. A neoprene wetsuit is a good way to stay warm above water or below. Wetsuits come in thicknesses from $\frac{1}{16}$ inch to $\frac{3}{8}$ inch and in a variety of styles, from full pants and jackets to vests, hoods, gloves, and boots. While chest waders will suffice if you don't intend to wade above your waist, in cool weather you may be more comfortable in a wetsuit top. If you'll be snorkeling or diving, a wetsuit is a must even if the water is warm. You won't be moving very fast, so you won't be generating much body heat. Even water 75 or 80 degrees Fahrenheit can feel cold after a half-hour of hunting. Hunters who work swimming beaches and piers with scuba gear often get four hours on one tank of air. If you'll be doing this kind of work, even in reasonably warm water, a $\frac{1}{4}$-inch thick suit would be a wise investment. Your local dive store can recommend a good suit based on your needs. Custom wet suits, tailored to your exact measurements, are more expensive than off-the-rack models, but the difference in effectiveness—particularly if you'll be doing long dives in cool water—is significant. To get the most out of your suit, wear a pair of coveralls over it. The denim will protect the neoprene from abrasion as you scramble around the bottom digging targets.

If you'll be hunting in water below 60 degrees Fahrenheit, consider buying a drysuit. Drysuits keep you completely dry. With thermal underwear or jeans and a sweatshirt underneath, you'll stay toasty warm in even the coldest water. Many waders use drysuits as well, finding them the best possible way to stay warm in the water.

Shown here are some of the options available to divers when a wetsuit is custom-made.

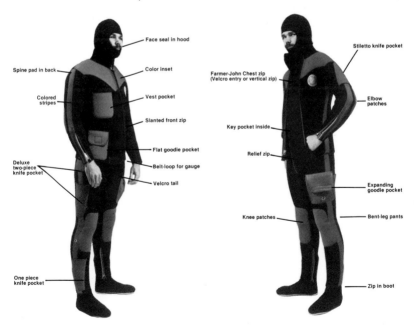

Retriever/Scoop. Retrievers or scoops are most useful for wading. While there are plenty of commercially-made models available, most hunters prefer to make their own. A retriever is really a combination shovel/sieve, so the handle should be long and the whole unit strong enough to stand up to serious digging. Making a substantial retriever is simple. Have a machine shop or a friend with a welder and a drill press help you. The basket on the bottom can be made of heavy-gauge steel or aluminum plate. Drill the entire plate with ½-inch holes one inch apart on center. Roll the plate into a cylinder and weld the seam. Close one end of the cylinder with a flat plate and shape the other into a shovel point. Your handle can be made of aluminum, steel bar stock, or electrical conduit. It's best to use two pieces of bar or conduit of unequal length. Attach the shorter piece to the top of the cylinder and the longer one to the back. Bring the pieces up into a single stem and bolt together. Bend the end of the longer of the two pieces into a flat loop to create a handle.

Divers generally use their hands—covered with tough work gloves—to fan the bottom away. However, you will occasionally find a target buried in a hard-packed mud or clay bottom. Most divers carry some type of digging implement, such as a scoop. A feed scoop, drilled with holes to allow silt and water to wash through, works well. More useful is a small gardening hand shovel or the blade of a hoe, attached to a short handle. Before diving with one of these, attach a loop of yellow polypropylene rope to the handle. The loop can then be attached to a D-ring clip on your weight belt to keep the item handy. Also, if you lay the shovel or hoe aside, the rope will float straight up, making it easier to locate in dark or silty water.

A number of divers use a suction dredge instead of digging by hand. The advantage is that the dredge can move the bottom pretty quickly, allowing you to search more ground more thoroughly. A dredge is a suction pump that sits on the surface. Its intake is through a hose under the water. A five-inch dredge is usually sufficient for most light treasure hunting work. The material sucked in by the pump is spewed out onto a sieve, where rocks and—hopefully—treasure collect. Usually, a dredge is most effective if one hunter works the bottom while another watches the sieve and sorts the trash from the treasure. Another method is to attach a wire basket just behind the head of the dredge. Heavy materials, like coins and jewelry, entering the dredge hose will fall down into the wire basket.

Sieve/Float. Waders often use a floating screen or sieve to sift their finds. An easy way to do this is to build a wooden frame and surround it with a motorcycle inner tube. An oblong frame of one by four, 15 ¾ inches by 19 inches, will work nicely. Sand the frame members and coat them with an epoxy-based paint, then nail together and cover the nail holes and joints with the epoxy paint. Nail or screw a piece of hardware cloth with a ½-inch mesh to the bottom of the frame. Put the frame inside of a motor-cycle inner tube and lash it on with nylon or polypropylene rope.

If you'll be snorkeling or diving in shallow water where there may be boat traffic, you should tow a float with a diving flag attached to it. If you'll be working a relatively small area and don't want to bother with hanging on to the float line, use a small brick or a lead diving weight as an anchor. Stay within five meters (15 feet) of the float, and as you move, move the float with you. The diving flag warns boaters to keep clear. The number of boaters who have no idea what the flag means is, unfortunately, rather high, so always look up at the surface, listen for boat engines, and check the complete 360 degrees before surfacing.

Goodie Bag. Okay, so you just found the twin of the Hope Diamond. How do you safekeep it until you reach land? Waders seem to prefer a heavy-weight denim pouch sewn with high-test monofilament fishing line and closed with a large-toothed plastic or nylon zipper. Denim seems to resist the effects of salt and water better than most fabrics. Don't trust anything you buy in the store. Sew your own, very carefully, and put the stitches very close together. Use a lockstitch and pull the thread tight. A nylon or plastic zipper won't rust, and the large-toothed ones seem to work well even when encrusted with salt. An occasional cleaning and treatment with silicone should keep it in fine shape.

Another alternative is to use a diver's game bag or "goodie bag." These nylon net bags come in several sizes. Most, however, are secured at the top with a very loose wire-handle-and-clasp arrangement. Consider buying a goodie bag and making a new "mouth" for it. Cut a rectangular piece of nylon cloth and sew a large-toothed nylon zipper into the center of it with monofilament. Then sew the edges of the goodie bag to the nylon cloth, using monofilament and a lockstitch. Attach a metal grommet to one corner of the cloth. A loop of polypropylene rope threaded through the grommet and attached to a D-ring clipped to your belt or wetsuit will keep the bag within reach.

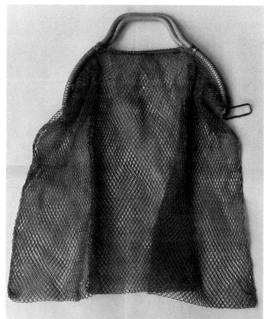

The mesh of a nylon net "goodie bag" such as this may be too wide to hold small objects. Be sure to use a ziplock bag inside if your finds are small enough to slip through the net.

Serious treasure hunters keep a log that includes information such as sites searched, weather conditions, and items found. Because notes are often jotted down near water, it's a good idea to use one of the water-proof or water-resistant log books available in dive stores.

Trash Bag. Another necessity is a bag to store pop-tops and other debris. Even the most experienced hunters say you have to work an area at least five times to get most of the treasure. The first time through, you'll be pulling up numerous bits of metal debris. Rather than have to find them again on your next sweep, stow them in a burlap or net bag and dispose of them properly topside. It's also good public relations for treasure hunting if folks see you removing trash from a beach rather than picking it up from the bottom and tossing it back in over your shoulder.

Markers. Coins and jewelry tend to accumulate in pockets. These may be at the juncture of several water currents or in a slight depression in the underlying bedrock. Where you find one object, you're likely to find others. If you hit a rich pocket, you may want to mark it, either permanently or temporarily. You'll be able to relocate it quickly if other parts of the area aren't running that day.

A temporary marker is simple to make. Paint a small piece of styrofoam yellow or orange and tie ten feet of thin polypropylene rope to it. Attach a lead diving weight or big fishing sinker to the end of the rope and wind it all up around the styrofoam. When you find a hot pocket, drop the weight, unwinding the rope.

A more permanent marker can be made with a styrofoam buoy marker. Most pool supply stores sell the oval styrofoam buoys used to mark off swimming areas. Thread one of those onto a piece of polypropylene rope and attach the rope to a brick or to a coffee can filled with cement. If you want to be able to find the area from above water, attach enough rope for the buoy to float on the surface. Surface buoys disappear, though, cut by boaters and carried off by waves. You may want to put on just enough rope for the styrofoam to float a meter or so (three feet) above the bottom. It'll be easy enough to find once you're swimming in the general vicinity.

Log Book. Over time, you'll find that some sites are more productive than others, and that certain weather conditions affect some areas more than others. The only way to keep track of this information—and use it to your advantage—is by keeping a log book. All serious treasure hunters keep a log. They note the area they are searching and identify the specific site by prominent landmarks (for example, "ten meters from shore, from the end of the jetty to the tip of the pier"); the weather conditions, including wind direction and velocity and the time of low- and high-tide; what was found and how deep in the bottom it was lodged; and any other important details. It's easy enough to keep a log in a spiral-bound notebook, but you may want to look into any one of the waterproof or water-resistant log books sold in dive stores, which withstand the wear and tear better than ordinary notebooks. Also, the standard information included on each page will make it easier to remember to write down the most pertinent details of location and weather. The most important consideration in using a log book is to be consistent. Always log your searches *before* you leave the area. Don't wait until you get home or you might forget.

Chapter 3

Where to Search

Treasure hunters often say that treasure is wherever you find it. More to the point, treasure of some sort can be found almost anywhere you look. Wherever people have been, they have left behind valuables—coins, jewelry, trinkets. Some of the items were lost, others deliberately thrown away. Yesterday's junk is literally today's treasure trove. Small lead toys fetch surprising prices from collectors as do common household items from the 17th and 18th centuries.

Most treasure hunters try to increase the value of their finds by hunting in areas likely to turn up antiques. Coins, among the most common finds, increase in value with age. Also, coins minted before 1964 were pure silver rather than the copper/silver sandwich used today. Gold coins were in common use up until about 1900 and the gold content of jewelry was generally higher prior to the 1960s. For all these reasons, treasure hunters seek out older swimming beaches, the sites of old resorts, and the like.

The key to good finds is good research. For every hour you plan to spend in the water, figure on spending two hours pinpointing the exact location of your search. Why are you searching in that spot? What do you hope to find? Where, exactly, you will find it?

Plowing through piles of musty documents sounds a lot less like adventure than pulling gold rings out of beach sand. But it's a necessary part of the adventure. Treasure hunters are more than people with metal detectors and time on their hands. Weekend coin-shooters can do well enough without in-depth research, but every professional treasure hunter is *first and foremost* an historical detective. They know where to look and what they're looking for before they dig up the first pop-top. The information that ultimately led treasure hunter Mel Fisher's crew to the *Atocha* and its half-billion dollar hoard came not from divers, but from historian Dr. Eugene Lyon. While researching his doctoral thesis in the Archive of the Indies in Seville, Spain, Lyon found an eyewitness account of the shipwreck. The actual site of the wreck was almost a hundred miles south of the area where Fisher—and others—had been searching for the *Atocha*.

Where Not To Search

Before detailing all of the places where you can find treasure, remember that you cannot search wherever you'd like. Private property, historic sites, and some public lands are off-limits to treasure hunters. Private property is private; hunt with the permission of the owner or stay out. It's that simple. Hunting on public land is a bit more complex. The "finders keepers" rule has been muddied by the long-simmering feud between archaeologists and treasure hunters that has its roots in the split between

academic and avocational historians that occurred in the 1930s. In the 1970s, that feud broke out into open warfare.

For the past decade, there has been an almost continuous uproar between treasure hunters and archaeologists over the ownership of antiquities. Some archaeologists believe that anything pre-dating John F. Kennedy's presidency is historic and belongs to the public without regard to the rights of whoever took the time and effort to find it. Treasure hunters, on the other hand, believe that anything they find should remain their property, to keep or sell as they see fit. Both of these positions are extreme. If you will profit from digging up the past, you have a responsibility to your fellow citizens not to destroy or unreasonably detain the artifacts that are part of their historic heritage. However, except where otherwise determined by the law, you have a right as a private citizen to look for, recover, and possess lost property. Especially when the lost property is located under water, the federal courts have consistently upheld the right of private citizens to excavate and recover artifacts.

This right is not unlimited, however. The federal courts have recognized the interest of the public in preserving historic sites and artifacts. They have applied a body of law known as "Admiralty Law" to the recovery of valuables. Admiralty Law is not a set of legal statutes, but rather a group of court decisions dealing with marine salvage which have accumulated over the centuries. A salvor may apply to a federal court for the salvage rights to a sunken vessel or submerged valuables. The court then "appoints" the salvor as legal custodian of the wreck or artifacts. The court protects the site from pilferage by other salvors—if they move in on the find, they can be cited for contempt of court and jailed indefinitely. The court also acts as an agent for the public, protecting its interests in the artifacts. The court's power to set regulations on the salvage of the site is virtually unlimited. As one archaeologist working the *Atocha* put it, "If the judge decided we ought to drink a beer after we recovered each coin, we'd have to drink a beer after we recovered each coin or face contempt charges."

Some states have specific laws governing the recovery of valuable and historic artifacts from public land. Public land includes, but isn't limited to: the bottom land underlying navigable waterways including rivers, streams, lakes, and the ocean (out to the three-mile territorial limit); historic sites owned by the state; state parks, forests, and game preserves; public boat landings and public beaches. Texas, in particular, has a very, very strong statute. The state-run Texas Antiquities Commission owns everything that comes up out of state waters. In general, the Commission is not interested in lost coins or jewelry, but if you were to hook into a nest of silver *reales* off one of the many Spanish treasure wrecks just south of Padre Island, you can bet they would come looking for you. Florida, which once had the toughest antiquities law in the country, lost ground in its court fight with Mel Fisher over the ownership of the *Atocha*. As a result, the state has backed off from its former hard-line stance against treasure hunting and is now actually working *with* salvors in the recovery of historic shipwrecks.

Most states do not have specific statutes governing the use of metal detectors on public land. Some of them have used existing statutes as the basis for regulations, however. It's up to you to determine whether there are restrictions that apply to the areas where you hunt. Most states have a state archaeological office. Generally it's known as the "shippo" (SHPO), or State Historic Preservation Office. Call the state capitol and talk to the people at this office. They may or may not be listed as SHPO in the telephone book. Try also listings for state archaeological survey, or call the main information number for the state's department of state, department of interior, department of recreation, or—if all else fails—the governor's office. Get a copy of the state's regulations and *be responsible*.

For more information on treasure hunting, historic preservation, and state laws, write:

**The Atlantic Alliance for
Marine Heritage Conservation**
Charles M. McKinney, Director
P.O. Box 27272
Central Station
Washington, D.C. 20038

**Federation of Metal Detector
and Archaeological Clubs**
Dick Stout, President
R.D. 2
Box 263
Frenchtown, N.J. 08825

The Atlantic Alliance, a non-profit organization, has been very active in fighting legislation aimed at curbing private treasure hunting. It sponsors educational seminars in marine salvage archaeology and acts as a national clearinghouse for information relating to treasure hunting and marine salvage. Membership in the Alliance is $20 per year for individuals and $10 for club members who join in groups of ten. The Alliance is an important lobbying organization for salvors and even operates a telephone hot line. You can call them at (202) 231-3666 with questions or comments.

Once you have secured permission to hunt a site and have familiarized yourself with state and local laws governing treasure hunting, you're ready to embark on the exhilarating quest for treasure. The "hot spots" listed below are merely a starting point. Do your research and be persistent in your search, and you're bound to uncover riches from any body of water that attracts people.

Water, Water Everywhere

Why search in the water? Why not confine yourself to land and leave wading and diving to the ducks? There are certainly a number of land sites that will prove productive for the hunter. But on an hour for hour basis, serious hunters say that water sites produce more and the finds can be recovered more easily.

Where there are people and water, the people will go into the water and successful treasure hunting will invariably follow. Any water-edge site that attracts people is an excellent prospect.

Swimming Beaches. For some reason, people swim wearing an amazing variety of expensive jewelry. The water shrinks their fingers, lubricating the goodies, and off those goodies go. Prior to the 1950s, people swam wearing a whole lot more clothes than they do today. Those clothes had pockets for change and other loose items. Often enough, the contents of the pockets ended up on the bottom.

Although some ocean beaches have been in continual use for a century or more, the popularity of some of the lesser-known areas waxes and wanes. This is especially true of freshwater beaches. Select beaches that have not been used since the early 1960s to eliminate the clad coins and cheaper modern jewelry. You'll most likely find a proportionally higher yield of antique jewelry at these sites as well.

Many abandoned freshwater swimming beaches were private recreation areas. The owners charged an admission fee and often operated concession stands near the water. These pay beaches attracted large crowds of people—and a fair weight of coins and jewelry.

If you have a college nearby, go to the library and look up some old yearbooks. Find out where the students swam and check the site for coins, class rings, and other jewelry.

Rope Swings. Rope swings are tailor-made for the treasure hunter. Here you have people swinging off a bank or out of a tree and hitting the water hard, maybe hard enough to dislodge a loose ring, jewelry, or coins. Look for "holes" in creeks or along canals or river banks—deep spots where swimmers might have dived. Look at the tree limbs overhanging the pool. A few loops of frayed rope wrapped around a limb marks the spot.

Canals. As a transportation system, canals enjoyed a relatively short, but prosperous, history. At the turn of the 19th century, canals sprang up along major rivers east of the Mississippi as a connecting link between rapidly-developing regions. Though the era of the canal ended with the development of the steam railroad, you'll undoubtedly find numerous reminders of the past. A few hours spent perusing historical records should direct you to the locks and perhaps the site of landings, popular inns, or the departure points for canal boat excursions attended by dignitaries, socialites, and politicians.

Docks and Piers. Docks and piers are a magnet for valuable goodies. In the late 19th and early 20th centuries, many lake and ocean resorts featured amusement piers lined with slot machines. Over the years, a lot of quarters and half-dollars went into the drink instead of into the slots. Often, the pilings or cribs of the piers can still be seen. Cribs were rectangular areas filled with rocks outside of the pilings. Filled with rock, they acted as a breakwater to keep storms from damaging the pilings.

Heavier objects can be found all around the cribs and at the base of the pilings. Be very careful working around docks and piers. In salt water, the pilings are often covered by razor-sharp barnacles that can slice and dice an unwary wader or diver thrown against them. Also, watch out for monofilament fishing line or fishing nets that may be tangled on the pilings. Whenever working around obstructions or in turbid water, carry a good scuba diver's diving knife with a serrated edge for cutting rope, cable, and monofilament.

Ocean Resorts. Everyone knows the legacy of Atlantic City, New Jersey, and Avalon, California. But there are dozens of other resorts—abandoned or still operating—from the Northwest coast all the way around to Florida and up the East Coast to Maine. At the very least, these resorts had a pier for fishing, relaxing, or boating. Most had designated sunning and swimming areas, which should prove rich in finds.

Water Parks. Amusement or recreation parks were common at lakes throughout the country until the mid-20th century. Lakeside beaches are an incredible storehouse of treasure. One treasure hunter has pulled over 200 Liberty half-dollars out of an amusement park beach that was abandoned about 30 years ago. His friends have recovered an additional several hundred half-dollars as well as more than 200 rings, a variety of charms, and amusement ride tokens.

Rafts. If a lake has a raft, there will very likely be treasure under it. Some older sites had platforms on pilings some distance from shore, and the area around them will most probably be littered with goodies. Inspect photographs of old swimming beaches or talk to elderly residents of the area to learn if there was a raft and—most important—where it was located.

Ferry Crossings/Boat Landings. Larger lakes and rivers may have had a ferry at some time. Find the landing points and you've probably found treasure. Because people on a ferry are usually fully dressed and doing business, landings are a likely spot to find higher-denomination coins such as silver dollars and gold pieces.

Stagecoach/Railroad Stops. Stagecoach or steam railway stops were often located near water, particularly canals and rivers. Passengers milling around during stopovers may have dropped valuables or even tossed pennies into the water for good luck.

Mills. An old mill site will usually yield a trove of interesting artifacts. Because many were abandoned in the 19th century, they may yield fewer pieces of jewelry and fewer coins. But the thrill of finding an 18th-century pewter plate or farm implement is indescribable. Many place names preserve the location of old mills. There are thousands of locales called "Milltown" or "Milton" as well as "Miller's Pond" and "Mill Hill." Find the stone piers that supported the mill wheel and you've found a good site.

Old Homesteads. Early settlers usually located their houses near a stream or another source of water. The stream bank or lake shore directly opposite the foundations of an old building will generally prove fruitful.

Fords. Before there were bridges or ferries, people crossed streams and rivers at shallow spots called *fords*. Again, place names are your clue. Many cities grew up at the sites of fords, and a "Redford," "Chadford," "Clifford," or other place with a name ending in ford and located on a river bank is probably such a site. Check the earliest maps you can find for the exact location of the ford. Remember when looking at old maps that most river beds change locations periodically. You may have to cross reference the early maps to more recent geological survey maps.

Jetties. Around the base of an ocean jetty you'll find not only the things fishermen have lost over the years, but a number of things that have washed down the beach and collected under the rocks. Remember, though, that jetties are usually positioned so as to catch the full force of the incoming waves and tides and prevent currents from filling a bay, boat channel, or river mouth with silt. The side of the jetty facing the prevailing winds is likely to be extremely rough much of the time. Be very cautious when working these areas.

Yacht Basins. Though most yacht basins or anchorages prohibit tenants from throwing items overboard, many goodies invariably end up in the water—ship's parts, tools, cookware, utensils, and the like. And, on occasion, watches, earrings, necklaces, pendants, and rings have made their way into the water too. You'd be surprised how many boat owners wear their valuables while swabbing the deck! Before you explore these water sites, be certain to ask permission of the boat basin manager. You may be able to locate abandoned yacht basins by checking old nautical charts or back issues of regional boating publications. Until recently, most boat fittings were solid brass. Collectors of nautical implements will pay plenty for a 19th-century sextant, or you may have a blank spot on your own wall for a fine marine instrument.

Bridges. For some reason, folks like to fish, sit, and otherwise pass time on bridges. Search around the pilings on the down-current side of the bridge or under rocks just downstream. Abandoned bridges can be found by referring to early nautical or geological survey maps of an area. While you may see nothing more than the old stone piers that supported the roadway above water, a number of valuable antiques may turn up below. If you are diving or wading at the base of a bridge, be careful not to get tangled in old fishing lines that may have snagged on the pilings.

Battle Grounds. Rivers and creeks are a natural line of defense, and many land battles of the Revolutionary War, Civil War, and various Indian and regional conflicts were fought on river banks. The Revolution and the Civil War left a fair number of ships on the bottom of lakes, rivers, and bays as well. Numerous volumes of military history are available, many written by soldiers who fought in the campaigns or the officers who commanded them. These are especially valuable if they include maps of the action. As a last resort, the Pentagon archives hold information about many of the battles fought by the U.S. Army and Navy, as do military libraries such as those at West Point, Virginia Military Institute, the Citadel, and the George C. Marshall Library at Carlisle University.

Research

The Library. The resources available to you for research are virtually unlimited. Most towns have a library. Get to know your librarian. He or she can be a bottomless well of inspiration and knowledge, helping you track down rare documents, early survey maps, and so on. Most counties have a branch library system in which valuable books and documents are kept in a research collection in the main library. Your local branch librarian should have access to a catalog of these items. Ordinarily, the librarian can request that the materials be sent from the central repository to your local branch where you can examine them at your leisure.

You may even be able to recruit your town's librarians as research assistants. Let them know what you're up to and what you're looking for. Tracking down rare maps, books, and documents will prove to be a lot of fun for many librarians. If their information results in a find, bring in your goodies. Let them see for themselves the fruits of their research.

Local Histories. Get to know the Dewey Decimal System yourself. The card catalog in your local branch library will list a number of books dealing with local history. You should read each of these thoroughly, making notes as you go, before bothering a librarian. These local histories will tell you the most astonishing things—the location of old stage coach stops, ferry landings, early inns or taverns, watermills, and recreational areas. You will amaze casual coin-shooters by pulling superb finds out of areas they've never even considered hunting.

Newspapers. Most libraries carry copies of the local newspaper on microfilm or microfiche. Starting with a year in the early 1960s, scan the

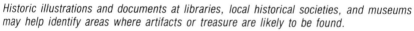

Historic illustrations and documents at libraries, local historical societies, and museums may help identify areas where artifacts or treasure are likely to be found.

spring, summer, and fall issues for references to beach concerts, swimming beaches, dance pavilions, popular fishing spots, fishing piers, amusement parks, boating areas, and so on. Check the interior pages, particularly the "social notes" column and the calendar of upcoming events, both of which refer to local events, the outdoor and sports pages, and the lost and found section of the classified ads. An ad such as "Lost at Bowman's Landing ... a diamond brooch in the shape of a heart ... Reward" tells you that there was once a place called Bowman's Landing and that something valuable was lost there.

Museums. Many towns have a museum that preserves the area's history. The collection of research material will likely be very thorough. The museum personnel may also be helpful in identifying and placing value on finds of historic significance.

County Courthouse. Your county courthouse holds a wealth of information. First, there are plat maps of the county. Plat maps are detailed, section-by-section maps of the area. There should be an archive of these going back to when the county was first surveyed. The mythical "Bowman's Landing" previously cited would be likely to show up on a plat map. The locations of old roads and bridges will be shown, and other features such as houses, watermills, docks and piers, and other man-made "improvements" may be plotted as well. A quick trip to the courthouse to examine a map of the mid-1950s and one from the 1850s will tell you how meticulous the county's surveyors have been.

Department of Parks and Recreation. The agency that maintains the county's parks should have files that detail all of the public recreation areas and may have information on some private ones as well. Their files can be very valuable because they will probably include maps and site drawings for docks, piers, boat landings, swimming beaches, and similar sites, many of which are no longer in use. Without their drawings, you'd be guessing as to the limits of a beach area or the exact location of a pier. Also, people who work for the department—particularly older employees who have been there for 20 years or more—can give you tips on abandoned public and private recreation areas. Some counties have regulations that restrict the use of metal detectors in county parks. A quick visit will bring you up to date on what rules apply and will perhaps steer you to valuable locations where you are allowed to search.

Health and Sanitation Department. While not a primary resource for most treasure hunters, the local department of health has the responsibility of overseeing the safety of recreation areas and food concessions. If you find references to a private recreation area in a newspaper or another source, but can't find its location on county survey maps or through the parks and recreation files, try the department of health. If any food was served at the area, a sanitation permit should be on file for each year it was in operation. The permits should give an address, which could then be located on the survey maps. Also, the details of the inspectors' reports may be on file and may indicate how large a crowd the place attracted.

Geological Survey Maps. Each state maintains a geological survey department, which is usually located in the state capitol or at a state university. The department will have archives that include the very earliest maps made of an area right up through current maps. Particularly when looking for river ferryboat landings and such, their files will be crucial. While you may find plenty of references to the location of a landing (such as 3.2 miles above a prominent bridge or below the mouth of a certain creek) the river bed may very well have changed over the years. The survey maps will show this clearly. The state's surveyors are always in the field, traipsing over areas where very few others go. If you are serious

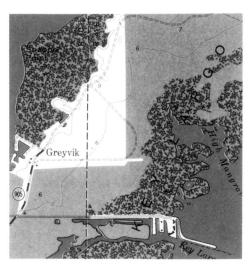

Mapped in 1949

Geological survey maps, which are revised periodically, show changes that occur in both the man-made and natural features of an area. Here, four maps of the same area in Key Largo, Florida, show that significant changes took place between 1949 and 1973. Such changes are likely to affect where a treasure hunter searches.

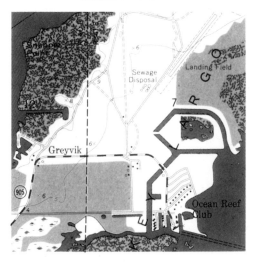

Revised in 1956

about treasure hunting, it would be worth your while to get to know one or two surveyors. The geological survey department will also have information on kinds of bedrock formations and bottom covering for many lakes, rivers, and streams.

In addition to state survey maps, you can make use of an excellent series of maps maintained by the U.S. Geological Service, which cover nearly every square inch of the country. Stores that sell camping and hiking gear often carry a set of the U.S.G.S. maps for nearby areas, or you can write for the U.S.G.S. catalog to: U.S. Geological Survey, 1951 Constitution Avenue, Washington, DC 20001.

Photorevised in 1969

Photorevised in 1973

Coast Guard. Maps of navigable lakes, rivers, and coastal ocean areas are maintained by the Coast Guard, and these can be purchased at any marine supply store. They indicate bottom formations, shipping channels, jetties, docks, piers, and shipwrecks, all good places to start your hunt.

Historical Society. All 50 states and most communities have historical societies. If such a group has a library, find out how you can get access to it (usually by joining). Historical societies also may have published or have available for sale books about an area that tell when it was settled and where to search for old homesteads, abandoned bridges, old recreation areas, and so on. Join your state historical society and attend the meetings. You'll be able to talk with others who are interested in the past, and some of them will have done extensive research on sites of interest to you.

Be forewarned, however, that not all members of the historical society will welcome you with open arms. In the past, treasure hunters have been deemed crass "pot hunters" who destroy historic sites in search of trinkets. This prejudice on the part of historians and especially archaeologists is not without foundation in fact. Early treasure hunters in Florida sometimes dynamited Spanish galleon wrecks to get a few handfuls of silver *reales*. Do not hunt truly historic sites that are being excavated by historians or archaeologists. Be prudent in your selection of sites and, when hunting, always be aware that you have a responsibility to the rest of the public not to despoil the area for others.

You may be able to use your hobby to ingratiate yourself with archaeological enthusiasts. Most of them spend more time reading about history than actually digging it up. You'll be in the field constantly. While hunting down relics, you may very well find things of interest to archaeologists—Colonial-period coins or artifacts, Indian relics, military equipment. Every community college has a professor whose hobby is local history. A nearby university is likely to have a department of anthropology or archaeology. The state archaeologists may have field offices near you. Cultivate relationships with these people. If you find anything they might be interested in, send them a photo or description of the item, or better yet, take it to them. They'll be able to tell you far more about the object than you could learn on your own—what it is, its probable date of manufacture, and what it was used for. They may be able to make suggestions about the nature of the site, its layout, and its use that will make your next trip more productive. Also, you may spark them to investigate some of your sites themselves. Putting a site under the protection of state historic statutes takes a lot of time and energy. Even if they thought your site had historic potential, it's doubtful that they would try to close it to treasure hunters.

Private Documents. Let your acquaintances—particularly those whose families have lived in your area for several generations—know what you do. It's likely that some of them have inherited letters, notebooks, or maps that describe potential treasure sites. Even an old homestead can be productive. A small farmer's rainy-day fund of 40 or 50 silver

Local historical societies often publish or make available information on sites that may yield impressive finds for treasure hunters. Historic papers or drawings showing warships under siege may identify underwater areas rich in artifacts of battle.

dollars, buried under a porch step, could bring you an excellent return on the time invested in hunting the site.

Treasure Hunting Magazines. Read the treasure hunting magazines regularly. They frequently feature successful treasure hunters. The specific areas mentioned may prove fruitful and you'll certainly learn how others work different types of sites.

The Corner Cafe. If you are working an area unfamiliar to you, stop at the local cafe before getting into the water. It's easy enough to start a conversation with other customers, and when you find one who's lived in the area for a while, you will be likely to have a great source of information.

Older Residents. The best kind of local resident to interview is an older person who has lived in the area for several decades. He or she will probably have a trove of information on old swimming beaches, resort hotels, docks, and other productive sites. What's even better is that many older people have extra time and are very happy when someone takes an interest in them. Some will literally take you by the hand and lead you to an abandoned and unmarked spot that's loaded. One treasure hunter recounts how an older gentleman took him out to a local lake and pointed to the exact spot where a permanent raft structure had been located about 30 yards offshore. The raft had been put up on pilings, and had been the main attraction of this abandoned swimming beach for four or five decades. Diving underneath the raft, the hunter found dozens of standing Liberty halves, Barber dimes, wheat-back pennies, and an excellent selection of gold rings and other jewelry.

Chapter 4

Search Techniques: Bringing It All Back Home

With a quality detector in hand and a notebook primed with promising sites, you're ready for the final stage of your hunt—search and recovery. The techniques used to locate and recover finds are as varied as the individuals who use them, but most successful hunters have two things in common: They are methodical and thorough.

There are two broad categories of hunters: Those who wade and those who dive. While there are some advantages to wading—and plenty of treasure has been found by waders—on balance, most serious hunters prefer to dive or snorkel while searching in the water, regardless of how shallow it is. As one hunter put it, "Even in lakes in areas that are less than four feet deep, I still use a tank. Sometimes the tank is bobbing out of the water, but it gets me closer to the bottom. I can find treasure in areas that waders tell me have already been worked. I can read deeper in the bottom while diving, and I can see the targets as they come up. It all works out to more treasure at the end of the day."

Search slowly and systematically, moving the detector's search coil about 18 inches (45 centimeters) to either side of your line of search. Keep the coil parallel to the ground.

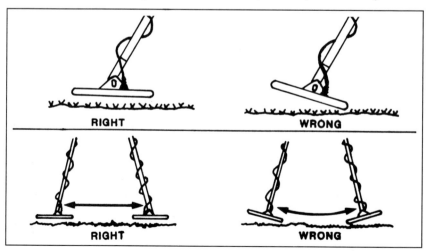

A Practice Run

Before going out to locate treasure with your detector, practice at home. It will make you more productive in the field. If possible, get some of the bottom material from the area you'll be searching. Bury a solid silver object, such as a pre-1964 quarter, about six inches deep in a clay pot full of the bottom material. In an area free of other signals (usually outdoors; inside, the nails in floor studs and wiring tend to interfere with the detector), sweep the coil back and forth across the pot. Listen carefully to the response in the detector earphones, as though you were listening to a song on the radio. Memorize the tone, the duration, the *sound quality* of the signal. A brash, clipped response, one that "hits" quickly, loud, and harsh, is probably a ferrous object, such as a nail or an aluminum pop-top. A smoother, more mellow tone is produced by silver or gold. It's difficult to characterize the sound of a treasure "hit." But with some practice, you will learn to distinguish gold and silver from aluminum or iron as quickly as you can tell Mozart from Michael Jackson.

In an area of your backyard that is relatively free of signals try burying coins, pop tops, nails, and other objects at various depths. Mark their locations with wooden or paper markers and run the detector over them. You might also try putting the objects underneath carpets indoors. Remember, though, that the nails in the flooring will give off strong signals. Practice walking slowly with the detector, making a methodical and thorough sweep of the area. Position the coil about four inches above the ground and begin to sweep methodically, as if you were operating a vacuum cleaner. Move the detector *slowly* from side to side, moving the

For units with variable "discrimination," such as the Fisher 1280-X Aquanaut, the sound of the signal is affected by the type of target being detected. The basic attributes of the signal are how quickly it rises and falls, whether it is brash or smooth, and whether the duration is short and peaked or longer with a slower fall-off.

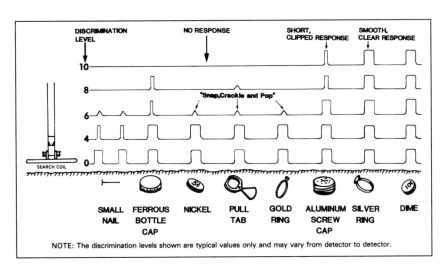

NOTE: The discrimination levels shown are typical values only and may vary from detector to detector.

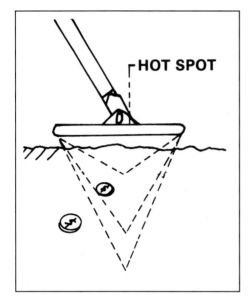

HOT SPOT

The area covered by the signal from a detector search coil is roughly conical and narrows the farther it gets from the coil. The signal "envelope" would be much larger for a large iron object, such as an anchor, than it would be for a small non-ferrous target like a coin or a ring.

coil about 45 centimeters (18 inches) to either side of your line of search. This will make the total width of your sweep about a meter (one yard) wide. Work in "lanes," much the way you would mow a lawn, and overlap the lanes by about 50 percent. If your sweep is a meter wide, position the centers of the lanes a half-meter (18 inches) apart. The coil penetrates the ground most deeply at the point directly under the center of the coil. Overlapping the search lanes will help you read faint signals from small gold objects buried deep in the ground.

Finding marked objects will also help you learn to pinpoint a target. It's not efficient to dig a hole the size of the search coil each time you get a signal. Work the detector back and forth across the buried objects. Listen for changes in the tone of the audio signal or meter reading as the coil approaches, passes over, and moves beyond the object. There are several means of pinpointing a target. One way is the "cross" method; you simply work the center of the search coil in an "X" pattern, moving the coil until the strongest signal is at the intersection of the two arms of the "X." That should be the location of the target. Another method is the "swirl." Move the coil in a circle around the area that produced the signal. Decrease the diameter of the circle until the coil is hovering right over the strongest signal.

A few hours of practice with known objects will make you a lot more adept when you first hunt in the water.

Wading Search Techniques

Waders do have a few advantages over divers. Waders have less trouble keeping their bearing during the search than divers working in murky water. Waders are usually able to cover an area more quickly—though not necessarily more thoroughly—than divers. It's easier to stay warm if you're not completely submerged, of course, and waders have less gear to haul than divers.

Gearing Up. Extend the wand or boom of your metal detector to a length that will allow you to hold the bottom of the coil about two inches above the bottom. It may be more comfortable to mount the electronics box of the detector on a belt worn around your waist. If the water is cold, make sure you're well insulated: Use either a wetsuit or a drysuit, or wear thermal underwear and thick pants (wool is best) under your waders, several layers of clothing on top, and heavy wool gloves or mittens covered with elbow-length rubber gloves of the type worn by chemical workers. Also, attach to your belt a "goodie" bag and a bag in which to store trash. Pick up your retriever, and your float if you'll be using one, and you're ready.

Start Walking. Once you have found an area you believe will produce, it's best to approach it systematically. A systematic search won't always produce more treasure, but that's the best bet. Define the limits of the area you'll be searching. Mark the beginning and ending points of your sweeps either in relation to natural landmarks or by putting up your own reference points. A pair of orange traffic cones set up in the sand are adequate. Once you have outlined the area, start at the high water mark. If you are working an ocean beach, search three hours before or three hours after high tide, and start your search at the high tide line. In a lake, start your search at the high water mark, which may be some distance above the actual water line, especially after a drought or if the lake is frozen.

Move slowly and steadily down the beach, keeping track of the beginning of each search lane and moving over only half the width of a lane on each sweep. When you get a signal, stop and pinpoint the object, then dig. Dig *each* target as you go, putting the trash in your trash bag so you don't have to find it again on your next sweep of the area. (Trash items close to the surface can "shadow" a valuable object beneath them. The signal of a pop-top or bottle cap close to the surface will mask the fainter signal of a silver coin or ring below it. You wouldn't be the first to find a valuable ring tucked inside of a bottle cap!)

Dig We Must. Some hunters find it difficult to position their retrievers due to the *refraction* caused by the water. When viewed from above the water, objects below the surface are magnified about 30 percent, and they appear to be displaced slightly from their actual location. You can prove this to yourself by dropping a pencil into a clear glass partly filled with water. The portion of the pencil under the water will appear larger and offset slightly from the part protruding above the water. If you pinpoint a target visually and then move in to dig, you are likely to dig on the wrong spot and come up empty. One popular technique is to use your foot as a guide. When you pinpoint the target, rotate your left foot sideways so that your toes are parallel with your shoulders. Place the heel of your left foot right behind the coil and in line with the center of the coil. Then move the detector coil away and rotate your left foot straight ahead without lifting it from the bottom. Your target should now be right under the ball of your left foot. This technique also works well when you're in dark or deep water and can't see the coil. If you can't see the coil, center your search pattern directly in front of one foot. That will give you a pretty good idea of where the coil is when you get a signal.

Thrust the retriever in firmly and deeply. If the bottom is firm, use one foot to force the retriever blade in; you want to scoop under the target and lift it out. Quickly bring the retriever basket up. If you're using a floating sieve, dump the basket in the sieve and check it for finds. Whether a find is in the basket or not, pass the detector over the hole made by the retriever again. If you didn't get the object on your first scoop, it may have been displaced slightly to one side by the retriever. Also, if you retrieved a piece of trash, it's very possible there's another target in the hole. Treasure tends to accumulate in "pockets"—that is, there are certain bottom formations that, due to their shape or to the prevailing wind or water currents, trap objects as they are moved around the area by waves and currents. Never move away from a hole until you've passed the detector over it again very carefully to ensure that you've cleaned it out. The first "hit" in one of these holes is likely to be on a piece of trash—a nail or a pop-top—because the signals they echo are stronger and louder than signals echoed by silver coins or gold rings.

When you've retrieved your target and the hole reads empty, immediately circle the detector around the hole. Something may have been displaced slightly when you pushed the retriever in. Also, since objects tend to accumulate in pockets, you may have dug on the edge or in the center of a pocket.

Cold Weather. If you will be wading when the air temperature is under 40 degrees Fahrenheit, you'll need to protect the batteries in your detector. Cold temperatures limit the amount of electricity produced by all types of batteries. Most waders remove the electronics package from the shaft, wrap it in quarter-inch-thick neoprene, and tuck it inside their waders. Your body heat will keep the batteries warm enough to remain charged.

Diving Search Techniques

Many serious treasure hunters feel that they're more productive when they're under the surface, rather than wading. By diving, you'll be able to get right down on the bottom where you can see the area you're searching, see where the signals are coming from, and see the hole clearly as you're recovering targets. There are three ways to accomplish that. You can snorkel, you can use a hookah rig, or you can dive with scuba gear. Of these three, using scuba equipment is probably the most convenient, but it does require that you take a course sanctioned by one of the national scuba certification agencies. Some treasure hunters prefer to work with a hookah, which does not require certification.

Snorkeling. Snorkeling is a very simple skill to master. The gear is minimal; all you need are a mask, fins, and a snorkel. The gear sold in discount chain stores is not adequate. Go to a professional dive shop and let them help you select the right equipment. The masks and fins sold by these shops are more expensive, but with good reason. They sell safe, dependable equipment made with high-quality materials. Expect to pay as much as $65 for masks, fins, and a snorkel. If you pay less than $30 you're either getting an incredible deal or you're buying second-quality goods. The manufacturers of quality gear include Cressi-Sub, Dacor, Farrallon-Oceanic, Nemrod, Plana, SeaTec, Scuba-Pro, Tekna, and U.S. Divers. There are also other reputable manufacturers. The professional dive store is your best source of information. You may also want to read the book *The Joy of Snorkeling*, published by Pisces Books, which describes snorkeling techniques in great detail.

With mask and snorkel, you'll be able to get down to the bottom to work the detector, but you won't be able to stay down more than a few minutes. Also, it's not a good idea to wear a weight belt while snorkeling, so it'll be hard to remain stationary long enough to dig up a target. Snorkeling is really practical only in water shallow enough so that your head is above water when you kneel on the bottom.

Quality equipment is a must if you intend to snorkel during your searches. The mask should be either high-grade silicone or real rubber, with a feathered inner seal and a tempered faceplate. Your local professional dive store can offer a selection of excellent gear and help you in selecting pieces that fit your needs and your body.

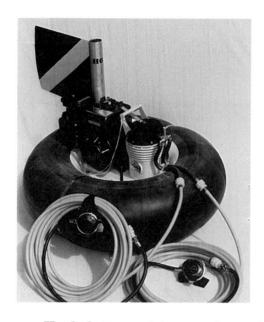

Use of a hookah, which pumps air from the surface to a diver below, will enable a treasure hunter to spend more time on the bottom than he would if snorkeling.

Hookah. For work in water deeper than a meter (three feet), you'll need either scuba tanks or a hookah.

A hookah rig is simply a gasoline-powered air compressor that floats on the surface on an inflatable platform. The compressor pumps air through a hose to a diver below. The diver may wear a conventional scuba mask and take the air through a mouthpiece, or may wear a "full face" mask—one that covers the nose and mouth. Full-face masks used for shallow water work are usually constant flow: The air from the compressor enters the mask continuously through an open valve, the diver breathes normally through his nose, and the excess air flowing through the mask is vented out into the water through a non-return valve. Many commercial divers, such as those who work on oil rigs, prefer hookahs to scuba gear because scuba gear generally cannot be used much below 40 meters (120 feet) and commercial divers usually work deeper than that. In fact, one of the divers who contributed to this book, Ron Houghton, made a world-record working dive below 400 meters (1,200 feet)! The U.S. Navy requires that any of its divers working below 30 meters (100 feet) use hookah rigs rather than scuba.

Hookahs are used frequently in shallow water, as well. There are no tanks to carry around, although the diver is tied to the surface by an air hose. This presents some difficulties when working around obstacles such as piers where the hose could become entangled or could be severed accidentally. The danger is minimal when working in shallow water because the diver can easily swim to the surface if the air supply is disrupted for any reason.

Unlike scuba gear, you can buy and use a hookah without instruction or certification of any kind. There are some fine points in using any type of diving gear, and an instruction course is the best way to learn them. Scuba courses are relatively inexpensive—usually around $100—and widely available. If you plan to use a hookah, you'll dive more efficiently and more safely after a certification course.

Diving and boating magazines usually carry ads from hookah manufacturers, and again, a professional dive store is an excellent source of information on the equipment. Some boat and recreation dealers may carry hookahs, but you'll be better served if you purchase one from a diving professional rather than a boating professional.

Scuba. *Scuba*, an acronym for "self-contained underwater breathing apparatus," is somewhat more convenient than a hookah. Rather than having an air hose trailing up to the surface, you carry your air supply in a metal tank on your back.

If you can swim, you can learn to scuba dive. You'll need to complete a certification course sanctioned by one of the national agencies: National Association of Skin Diving Schools (NASDS); National Association of

Experienced hunters say that scuba gear drastically increases the amount of treasure you'll find per hour of searching. If you're not already a certified diver, your local dive store is a good place to start exploring equipment and courses. Many local YMCAs and community colleges offer scuba courses as well, and some innovative schools now offer complete courses in two or three intensive sessions.

Underwater Instructors (NAUI); Professional Association of Diving Instructors (PADI); Scuba Schools International (SSI); or the Young Men's Christian Association (YMCA). It's possible to buy scuba gear without being certified, but you will not be able to get a dive store to fill your tanks with air unless you have a card proving that you have completed a course sanctioned by one of these agencies or a recognized foreign agency such as Confederation Mondiales Activities Subaquatique (CMAS) or the British Sub-Aqua Club (BSAC).

Local YMCAs, many community colleges, and dive stores also offer scuba courses. These six- to eight-week courses include classroom instruction and, in some programs, actual scuba instruction in a swimming pool. If you are in generally good physical condition you'll have no problem completing the course. While the Navy picks vigorous young men for its diving crews, it's not necessary to be young, a man, or even particularly vigorous to complete a scuba course. One of the most experienced divemasters in South Florida is a woman in her mid-50s who didn't learn to dive until she was more than 30 years old. After 20 years of diving, however, she is most certainly vigorous and has been known to outswim men half her age on the reefs and wrecks off West Palm Beach.

Once certified, you'll be able to get air tanks filled at any dive store in the U.S. or abroad. There's another incentive, too. Part of the money you make digging treasure out of the bottom of muddy lakes can be spent diving beautiful coral reefs in crystal-clear water when you visit the Caribbean, Micronesia, Polynesia, Australia, or the Red Sea.

Weights. Whether you are using a hookah or scuba gear, treasure diving does require some special techniques that aren't generally taught in a basic certification course. The human body is buoyant—it tends to float upward. When you're swimming on the surface, this is a good thing. But when you're trying to stay under, it's annoying. The solution is to don lead weights. In a scuba course, you'll learn how to calculate the exact amount of weight you need to carry to become neutrally buoyant so that you neither sink nor rise in the water. In treasure diving, however, you want to hug the bottom, to be able to dig into it and remain stationary. The solution is to carry extra weights. Although over-weighting is not recommended by any of the certification agencies, many working divers find it a necessity. The more weight, the more stable you'll be on the bottom. However, if you should have to ascend for any reason—a malfunction of your gear, for instance—the extra weights could be a problem. Some divers say that an extra ten kilograms (22 pounds) will help you stay put on the bottom. If you do wear extra weights, be certain that you wear them on a quick-release weight belt and that the belt will fall clear of your body and other gear instantaneously if you need to jettison the extra weight.

Another way to stay put on the bottom is to carry a small Danforth anchor attached to a length of rope. Especially in a heavy current, it can keep you stable without overweighting you. Tie three meters (ten feet) of nylon rope to the anchor. On the bottom, plant the anchor up-current from the area you'll be searching. Facing into the current, let the water push you back while you hold onto the rope. Just let out a little rope on each sweep. You might want to tie knots at intervals that correspond to the length of your detector's shaft. That way you don't have to judge how much rope to let out each time, you just back up to the next knot. If you use an anchor *be certain* that the rope is in *no way* attached to you or your diving gear. Also make sure that the rope doesn't become entangled in your gear while on the bottom. Keep a sharp diving knife with a serrated edge in a place where you can reach it easily. Traditionally, divers have carried knives in sheaths strapped to their calves, but if you get tangled in something, you may not be able to reach back to your leg. Better to carry the knife in a sheath strapped to your forearm. Consider this when selecting a knife.

Search and Recovery. One major difficulty of deep-water treasure hunting is that you don't have above-water landmarks to use as reference points. There are several ways around this. If you're searching near a structure, such as a pier, just keep in mind where you are in relation to the land end. Are you on the right hand or left hand side? Which direction are you facing, away from land or toward it? It's incredibly easy to get turned around while diving, especially in dark or silty water. You can also use submarine landmarks to guide your search. Find a depth contour, where the bottom abruptly drops a short distance, and swim along it in one direction. In freshwater lakes, you can search a weed line—where aquatic weeds start—which generally occur along a bottom contour or where the bottom changes from gravel to sand.

If there are no structures or other landmarks, you may have to carry your own to make a thorough search. The buoys described in Chapter 2 work well. Or, you can use white plastic jugs partly filled with gravel and

When snorkeling or diving, it's always a good idea to tow a buoy or a float with a diver's flag attached to it. The diver's flag will warn boaters of your location. A buoy with an anchor can be used as a reference point in turbid or dark water.

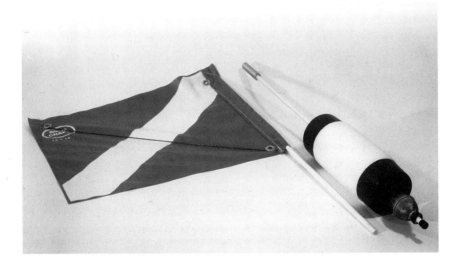

topped off with water. Tie a yellow nylon line between the handles of two jugs, set them six meters (20 feet) apart and use the line as a guide. By moving the jugs half the width of your sweep each time you make a complete run, you'll be able to blanket an area without missing any spots. You can use a knotted line tied to an anchor as described above to make circular sweeps. Just plant the anchor in the center of the area you want to search. Swim a complete circle, sweeping as you go. When you've completed 360 degrees, let yourself out one knot and circle around the anchor again.

Hearing your signals underwater can be a bit of a problem. Water is far from quiet. You'll have your own exhaust bubbles gurgling around your head. If there's any activity nearby, you'll hear that too. Boat engines can be heard at a remarkable distance underwater. It's also difficult to keep headphones secure while diving, particularly if you're in a strong current. One way to combat these problems is to remove the earphones from the headband and shove them up under a scuba hood. The neoprene hood will hold them tightly against your ears. *Do not* do this at the surface, however. Only put earphones over your ears after you're on the bottom. The earphones could trap air against your ears and could also interfere with the equalization of pressure between your inner ear and the water outside. This could lead to any number of painful and injurious consequences for your eardrums.

A dredge is probably the most effective way to work a very productive area thoroughly. Water entering the dredge hose pulls in sand and light objects with it, then lifts them to the surface. Heavier objects—such as coins—are then exposed.

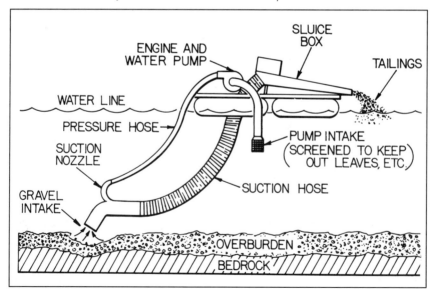

Dig This. Once you've hit a signal, stop where you are and dig the target. It's almost impossible to find a spot again once you've moved away from it. Divers tend to use their hands—protected by gardening gloves—instead of a retriever. If the bottom is sand, loose mud, or silt, all you need to do is slowly fan your hand back and forth. This will move the bottom sediment and allow you to watch for targets. If you're in a current, such as a stream, the silt will be carried away by the moving water, keeping the search area relatively clear. Be prepared to catch small objects, such as dimes, that flip up out of the hole and get caught up in the current. If you're working in calm waters, position yourself so that your chest is directly over your signal. Using one hand only, fan in one direction. Before bringing your hand back to its original position for the next stroke, turn it parallel to the bottom. This will set up a slight, steady current, moving the bottom sediment away from you and out of your face so you can see what's in the hole.

When working in loose sand or silt, be aware of your feet. A few careless kicks with your fins will stir up an instant Sahara-like sandstorm, limiting visibility for yards around. If there's a current, always face into it; silt will then be swept behind and away from you. Learn to keep your fins up off the bottom when moving around. If you're having problems with silt, try moving forward by sticking your fingers into the loose silt and dragging your body forward without kicking. Cave divers use this technique, called *finger walking*, to avoid silting out the narrow passages of a cave.

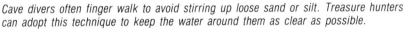

Cave divers often finger walk to avoid stirring up loose sand or silt. Treasure hunters can adopt this technique to keep the water around them as clear as possible.

If you're working in hard-packed mud or clay, you may have to use a small gardening hand shovel or the head of a hoe attached to a short handle to loosen the sediment. Then, fan as directed above.

If the bottom is rocky, you'll need to move some of the smaller rocks out of the way to get down to sediment. Rocky bottoms are almost impossible to work while wading, but divers find them very productive. Rings and coins—unidentifiable by a faint signal—tend to get caught under the rocks.

Don't leave a hole after finding the first target until you've swept it again with the detector to ensure there's nothing else in there. Also, as you fan or dig, you'll create a mound of sediment off to one side of your hole. Pass the detector over it carefully. You may find a small target that slipped through your hands or was pushed up over the edge of the hole. Again, treasure tends to accumulate in pockets, so that the first find may be no more than an appetizer. One diver says he's spent four hours working an area no more than three meters (ten feet) in diameter. If a hole is producing, keep fanning away at the edges, enlarging the hole, and making it deeper. The deeper you go, the more likely you are to find antique jewelry and pure silver coins. Don't be surprised if, after a half-hour, you're sitting down in the center of a hole a meter or more deep!

As was mentioned in Chapter 2, the best way to work an area quickly and thoroughly is to use a small suction dredge, one with a ten to twelve centimeter (four to five inch) inlet hose.

Chapter 5

Maintenance and Safety

Any kind of treasure hunting—especially if done in the water—poses some hazards. You'll be poking about in unfamiliar surroundings, perhaps around or in abandoned structures, perhaps some distance from telephones and medical facilities. Be careful.

It's always a good idea to hunt with at least one other person. Diving instructors strongly recommend that you do not dive alone and that you keep your partner in sight at all times. Even if your "buddy" does no more than watch you hunt, you'll be safer if there's someone around in case of trouble.

It may sound silly, but people who spend time in the outdoors know that the easiest way to stay out of trouble is not to get into any. Use common sense. The inside of a wooden barge that's been beached for several decades will offer plenty of opportunities to get trapped underneath rotted timbers or cut by sharp edges. Old diving piers can be just as dangerous, especially if the beams and supports are rotted. Wading around in water so dark that you can't see the bottom is generally a bad idea. Don't become so absorbed in your hunt that you fail to notice that the subsurface terrain has changed, that it's getting dark, or that you're wandering into potentially dangerous territory. Stay alert to your surroundings.

Protect your hands and feet from glass, sharp metal, nails, and other hazards. Wear heavy leather gloves if you'll be running your hands through the bottom sediment. Wear tough-soled shoes if you'll be wading. One way to avoid stepping on something sharp or into a hole is to shuffle your feet as you wade, dragging them across the bottom rather than picking them up as you move forward. If the bottom is thickly cobbled with slick rocks, be certain of your footing before bearing down with your full weight.

First Aid for You. Carry a first-aid kit with you at all times. It should, at least, contain bandages, large adhesive pads, gauze, surgical tape, scissors, a splint, an arm sling, a tourniquet, alcohol, antiseptic cream, burn ointment, sun screen, and a first-aid guide. Well-stocked, compact kits can be purchased at outdoors supply stores, dive stores, and boating stores. Get one with a watertight lid that seals with a gasket.

In addition, if you'll be in a truly remote area, you might want to take along survival supplies such as matches, a compass, fishing line and hooks, freeze-dried food, a flashlight with fresh batteries, a compact "space blanket" that conserves body heat, and a survival guide.

First Aid for Your Gear. What about care of your detector? Even the venerable Nikonos underwater cameras, designed to operate at depths to 60 meters (200 feet), occasionally flood. Just as you carry a first-aid kit for yourself, carry one for your gear.

If the electronic components in your detector flood, the primary damage will be done. But you can minimize the extent of the damage and prevent further damage by immediately drying out the detector. First, disconnect the power source. Turn off the detector (even if it has apparently stopped operating, turn the power switch off). Try to vent the affected area and drain the water out. *Very carefully* attempt to remove the power source. Don't stand in waist-deep water and just reach in and grab the batteries. Take the detector ashore. Dry yourself off and then try to work the batteries out with a pair of pliers with insulated handles, or isolate yourself from the electricity in some other way. After the batteries are out, assess the damage. Many amphibious detectors are designed with more than one watertight compartment. Determine which have leaked and what was in those compartments. You may find that you have nothing more than a wet battery compartment.

If the detector was flooded with salt water, rinse the affected area *only* in clean, fresh water. Salt is extremely corrosive and you'll do more harm leaving it on the electrical and metal components than you will by rinsing it off. When the affected areas have been cleaned, flood them again with alcohol. The small amount of water left in those areas will be diluted in the alcohol. When you dump out the alcohol, the affected areas will dry very quickly as the alcohol evaporates. The less time moisture is inside the detector, the less damage it will do. The best thing to do next is to take the detector to a dealer or a service center. If you feel lucky and the flood was definitely confined to the battery compartment, you might try inserting fresh batteries (*not* the wet ones) and turning the detector on. If you were right, it will probably work well. If you were wrong, the entire unit will probably short out when fresh juice surges through the still-damp components.

In any event, you'll need to determine the cause of the flood. First, look at the joints, compartment covers, and through-case fittings (such as electrical connectors and control shafts). Look for cracks, pits, or other evidence of failure. If there are compartment covers that fasten on, look at the O-ring seals that are supposed to keep them watertight. Are the O-rings properly seated in their grooves? Has dirt or grit accumulated on any O-rings? Is any O-ring just slightly slick with silicone grease or has it dried out? Check the fasteners: If they are snaps, are they still tight or have they loosened up with age? If they are screws, are the threads stripped or is rust preventing the caps or the nuts from seating firmly and tightly against the case? If you can't determine the cause of a flood, take the detector to a dealer or a service center before taking it back into the water.

The best way to prevent floods is to check the detector thoroughly before and after each use. As soon as you're out of the water, make sure that the power is off. Rinse the detector in clean, fresh water and dry it thoroughly with a towel. Take care to wipe down the areas under the knobs, around the control shafts, and around electrical connectors. If there are compartment covers, take them off and clean the O-ring seals and their seats. Take out each O-ring and check it for grit and abrasions. If the O-ring feels dry, grease it slightly with O-ring lubricant. This silicone preparation can be found in dive stores. Take the O-ring out of its groove. Rub a small dab of silicone grease between your thumb and forefinger, then pull the O-ring through your thumb and forefinger. You don't want a lot of grease on the ring; too much lubricant can cause a ring to slip under pressure and actually create a leak. Replace the O-ring if it appears worn or nicked. Check the groove in which the O-ring sits and the seal facing it for cracks or nicks. If the groove or seal shows excessive wear, take it to a service center. Occasionally, lube the connectors, electrical cables, metal fittings, and any fastener threads with silicone grease as well. It's an excellent preservative and prevents rust.

Your gear should include a fishing tackle box containing a set of jeweler's screwdrivers, needle-nosed pliers with insulated handles, a spare O-ring for your detector, silicone O-ring grease, a quart of alcohol, and fresh spare batteries. If you plan to dive with scuba gear, carry spare straps for your mask and fins, extra O-rings for the scuba tank valve, and an adjustable crescent wrench for the fittings on the scuba regulator. If you have a hookah, you may also need spare fittings for the hoses, a fresh spark plug for the engine, extra filters for the intake manifold, and an inner-tube repair kit in case the inflatable platform is accidentally punctured.

Appendix

Whether you're an aspiring professional underwater treasure hunter or a weekend hobbyist, you'll want to be well-equipped and well-educated about underwater exploration. The organizations and publications listed in this section are just a sampling of the many informational sources that can help you buy the right equipment—particularly metal detectors—and teach you more about the exhilarating world of underwater treasure hunting.

U.S. Underwater Metal Detector Manufacturers

Fisher Research Laboratory
1005 "I" St.
Los Banos, CA 93635
(209) 826–3002

Garrett Metal Detectors
2814 National Dr.
Garland, TX 75041
(214) 278–6151

Intex, Inc.
4720–P Boston Way
Lanham, MD 20706
(301) 731–4545

J.K. Gilbert Co.
700 Hwy. 80 S.
Leesville, TX 78112
(512) 424–3409

J.W. Fishers Mfg. Co.
Anthony St.
Taunton, MA 02780
(617) 822–7330

Turtle Corp.
51495 S.R. 19
Elkhart, IN 46514
(219) 262–4890

White's Electronics, Inc.
1011 Pleasant Valley Rd.
Sweet Home, OR 97386
(503) 367–2138

Government Agencies

If you're an historic shipwreck enthusiast, you'll find government agencies to be an invaluable source of information about the location, excavation, and preservation of shipwrecks. Some organizations can even help you locate a shipwreck if you can provide the date of sinking and its approximate location. One word of caution: some historic shipwrecks are protected by state and federal salvage laws, which may also prohibit the use of metal detectors. Investigate the law *before* you venture out, and do your part in preserving the cultural heritage of these shipwrecks.

National Archive and Records Service, National Archives, Washington, D.C. 20408. This service maintains records on ships wrecked between 1939 and 1974.

National Park Service, Submerged Cultural Resources Unit, P.O. Box 25287, Denver, CO 80225. The National Park Service maintains numerous underwater parks in the Continental U.S. and sponsors underwater searches and excavations for professional and amateur divers. NPS also publishes several newsletters and other materials of interest to underwater professionals.

United States House of Representatives Merchant Marine Fisheries Library, 550 House Annex-II, Washington, D.C. 20515. Upon request, the library will supply transcripts of hearings and other legislative action related to shipwreck archaeology and preservation.

Associations

Atlantic Alliance for Maritime Heritage Conservation, Inc., P.O. Box 27272, Central Station, Washington, D.C. 20038. Founded in 1983, this organization sponsors educational seminars in marine salvage archaeology and acts as a national clearinghouse for information related to treasure hunting and marine salvage. The Alliance is active in shipwreck legislation and is an enthusiastic lobbying group.

Federation of Metal Detector and Archaeological Clubs, R.D. 2, Box 263, Frenchtown, N.J. 08825.

Treasure Hunting Magazines

Western & Eastern Treasures, published by Peoples Publishing Co. Inc., 901 W. Victoria, B-2, Compton, CA 90220.

Lost Treasure, published by Lost Treasure, Inc., 15115 S. 76th E. Ave., Bixby, OK 74008.

Treasure, Treasure Search, and *Treasure Found*, all published by Jess Publishing Co. Inc., 6280 Adobe Rd., 29 Palms, CA 92277.